Jersey Geology Ramblers' Guide

Ralph Nichols Ph.D.

Jersey Geology Ramblers' Guide

First Published in Great Britain
in 2020 by Charonia Media

Illustration and design by Mark Jackson.

ISBN 978-0-9560655-7-5

Contents

Introduction

These trails are the result of discovering the striking variety of rock types along our coastline and exploring various outcrops around the many different coves, cliffs and cliff paths.

This variety was so much greater than in the deserts and plains of Australia and Canada where I mapped and interpreted various outcrops and studied drill cores.

In Jersey, there is much more colour and many more textures and structures both sedimentary igneous and tectonic.

Then I searched the publications on the subject and found only scholastic journal papers and university graduate publications which, of necessity were written in the technical language of the subject.

Thus, at one stroke, the ramblers, beach combers, tourists and local families and children were denied an introduction to the fascinating world of colour, texture and structure in the rocks around them, their history and the making of the island.

In addition, the action of the rain, wind and sun, the streams and the waves which have attacked the rocks and altered them and so changed the landscape which makes Jersey so attractive, were not described or evident to the passerby.

Jersey's geology differs greatly from that of the other islands and reefs and the surrounding sea floor, and from that of Normandy and Brittany.

There are sedimentary rocks from submarine deposition and other types from flash floods. There are igneous rocks, varieties of volcanic rocks formed from eruptions of different types of lava, of granites, diorites and dolerites, formed from magma below the surface, and even some metamorphic rocks, produced by heat from the granites.

These have been uplifted, folded and faulted at different times then eroded to form the island. We were then surrounded by warm seas in which chalk and later, fossiliferous limestone formed.

In the last 2 million years we were affected by the Ice Ages when the sea advanced and retreated several times leaving raised beaches, wind-blown glacial silt (loess) and head respectively. During the last 10,000 years, these were overlain by peat beds and sand dunes, and reworked to produce the beaches at present day sea level.

Visits to the nearby reefs and islands of Les Ecréhous to the north east and Les Minquiers to the south reveal low lying masses of varieties of granite which are closely fractured (foliated).

We are separated from them, and other sea - floor sedimentary rocks, by faults, which seem to make us an isolated block, a higher land remnant sitting on a raft of igneous rocks interpreted from geophysical studies. Some of these faults are tear faults which show Jersey to have moved from NE to SW, but not how far or from where.

Our oldest rocks named the Jersey Shale Formation, are dark grey shales interbedded with grey to red - brown sandstones (greywackes) and thin light grey, often laminated beds, called turbidites, which show lots of sedimentary structures. These beds vary from near - horizontal to vertical showing they have been uplifted and folded several times to form mountains.

The overlying volcanic rocks, are named respectively as the St. Saviour's Andesite and the St. John's & Bouley Rhyolite Formations. The andesites consist of dark grey vesicular lavas, ash and pyroclastic beds and , some with large white crystals, making them porphyritic. Above them are red - brown rhyolites, which various ash (ignimbrite) and spherulitic beds.

All the volcanic rocks have flow structures, flow bands and flow folds, and intervals of angular fragments (pyroclasts).

Then you can follow a granite trail along the cliff paths, because three granite masses intrude the above rocks in the NW, SW and SE of the island in which there are several varieties of granite and inliers of black and white speckled diorite and darker gabbro.

Our final, and uppermost hard rock formation, the Rozel Conglomerate Formation, occurs in the NE of Jersey and is a striking mixture of red sandstones and siltstones at the base, overlain by variously coloured pebble beds, of differing thickness, containing pebbles and rare boulders of the above rocks, including a granite not found on the island.

References.

Bishop, A. C. & Bisson, G. 1989. Classical areas of British Geology: Jersey: description of 1:25,000 Channel Islands Sheet 2. (London HMSO for British Geological Survey).

Jones, R. L., Keen, D. H., Birnie, J. F. & Waton, P. V. 1990. Past Landscapes of Jersey. Société Jersiaise.

Keen, D.H. 1978. The Pleistocene deposits of the Channel Islands. Rep. Inst. Geol. Sci. No. 78/26.

Mourant, A. E. 1932. The spherulitic rhyolites of Jersey. Mineral. Mag. Vol. 23. pp. 227 - 238.

Mourant, A. E. 1933. The geology of eastern Jersey. Q. J. Geol. Soc. London. Vol. 89. pp. 273 - 307.

Mourant, A. E. 1935. The raised beaches and other terraces of the Channel Islands. Geol. Mag., Vol. 70, pp. 58-66.

Nichols, R. A. H. and Hill, A. E. 2004. Jersey Geology Trail. Private publication; printed by The Charlesworth Group.

Renouf, J. T., & Andrews, M. 1996. Geology, pp. 17 - 22. In Les Écréhous, Rodwell, W. 1996. Société Jersiaise.

Renouf, J. T. & Bishop, A. C. 1971. The geology of Fort Regent road tunnel. Ann. Bull. Soc. Jers. Vol. 20. pp. 275 - 283.

Renouf, J. T. 1989. Rozel Conglomerate Formation. In Bishop. A. C. and Bisson, G. 1989. Classical areas of British Geology: Jersey: description of 1:25,000 Channel Islands Sheet 2. (London HMSO for British Geological Survey), p.40.

Salmon, S. 1996. Cylindrical granodiorite pipes in the Sorel Point Igneous Complex, Jersey, Channel Islands. Proceedings of the Ussher Society, 9 (1), pp. 114 - 120.

Salmon, S. 1998. The plutonic igneous complex at Sorel Point, Jersey, Channel Islands: a high level multi-magma assemblage. Geological Journal, 33, pp. 17 - 35.

Went, D. & Andrews, M. 1990. Post-Cadomian erosion, deposition and basin development in the Channel Islands and northern Brittany, pp. 293 - 304. In D'Lemos, R., Strachan, R. S. and Topley, C. G. (editors). The Cadomian Orogeny. Geol. Soc. Spec. Publ. No. 51. The Geological Society. London.

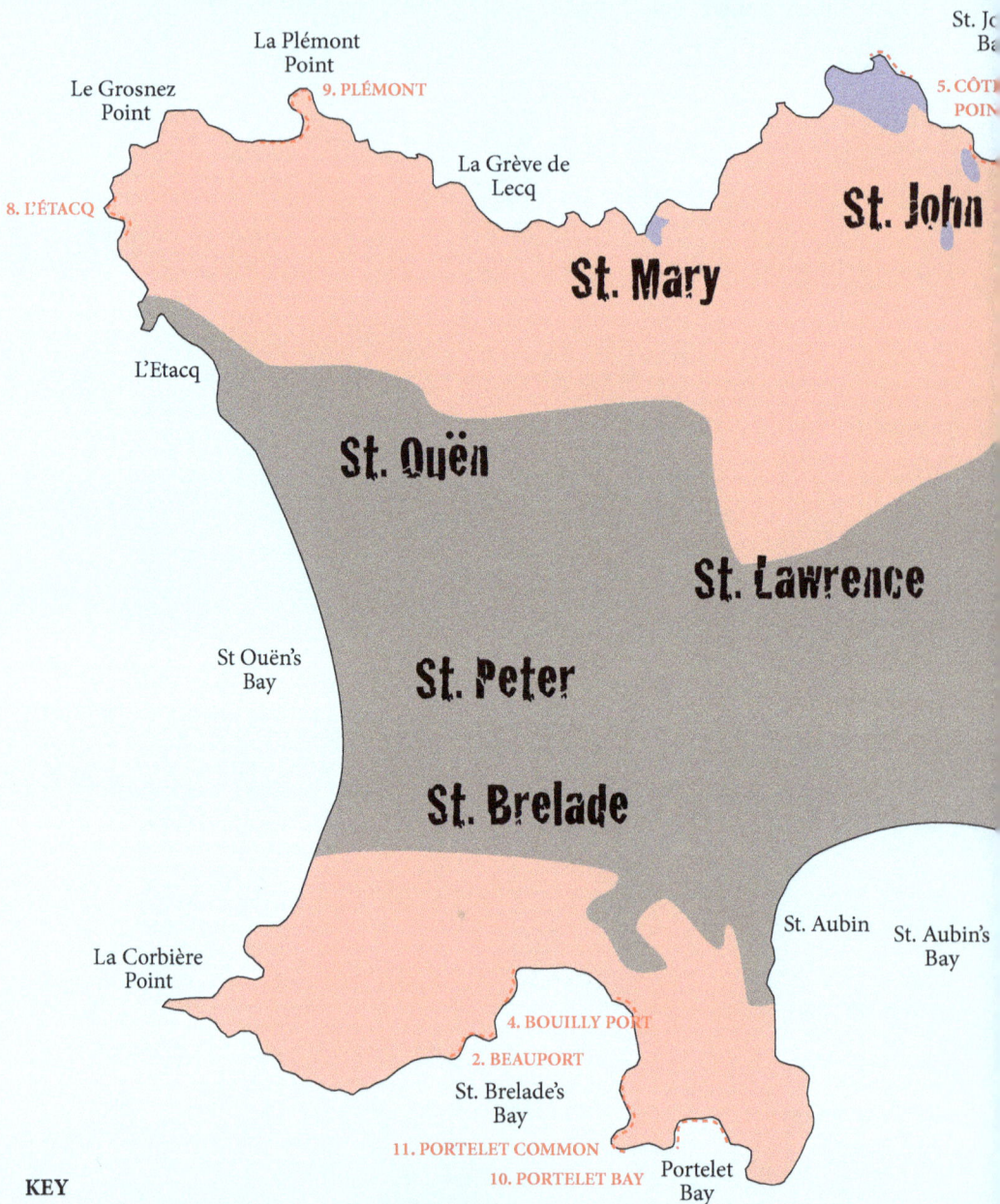

Le Grosnez Point

La Plémont Point

9. PLÉMONT

St. Jo
Ba

5. CÔTI
POIN

8. L'ÉTACQ

La Grève de Lecq

St. John

L'Etacq

St. Mary

St. Ouën

St. Lawrence

St Ouën's Bay

St. Peter

St. Brelade

La Corbière Point

St. Aubin

St. Aubin's Bay

4. BOUILLY PORT

2. BEAUPORT

11. PORTELET COMMON

10. PORTELET BAY

St. Brelade's Bay

Portelet Bay

KEY

Rhyolites	Jersey Shale	Geology Trails	
Rozel Conglomerate	Granites		
Andesite	Diorites		

Geology Map of Jersey

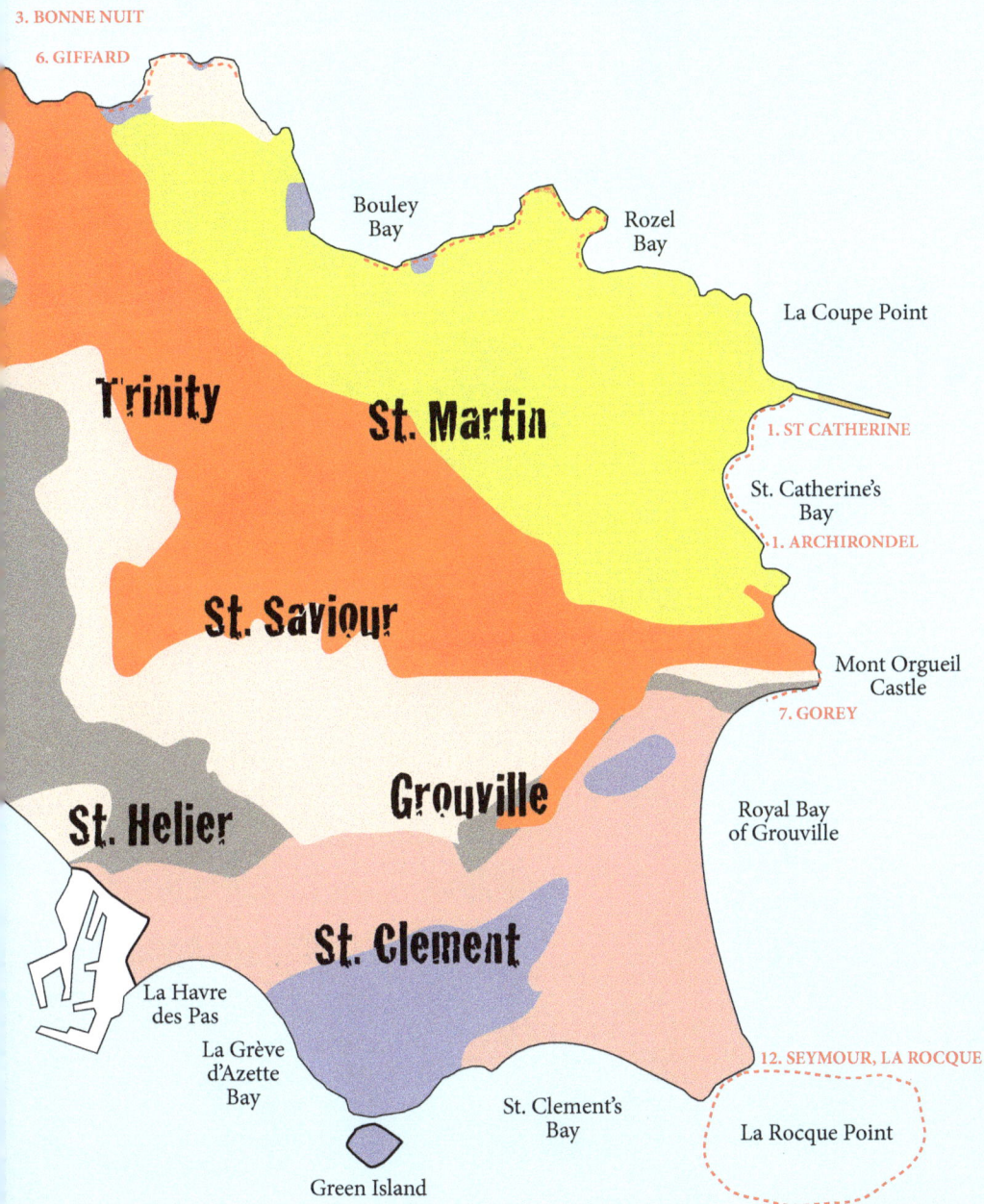

3. BONNE NUIT

6. GIFFARD

Bouley
Bay

Rozel
Bay

La Coupe Point

Trinity

St. Martin

1. ST CATHERINE

St. Catherine's
Bay

1. ARCHIRONDEL

St. Saviour

Mont Orgueil
Castle

7. GOREY

Grouville

Royal Bay
of Grouville

St. Helier

St. Clement

La Havre
des Pas

La Grève
d'Azette
Bay

12. SEYMOUR, LA ROCQUE

La Rocque Point

St. Clement's
Bay

Green Island

Did you know...

That there are three groups of rocks; Igneous from magma and lava, Sedimentary from deposition under water and air, and Metamorphic from heat and pressure?

The Archirondel – St. Catherine's Trail

Palaeozoic volcanic rocks, conglomerate, Pleistocene raised beaches, mid-tide grit & sandstone beds.

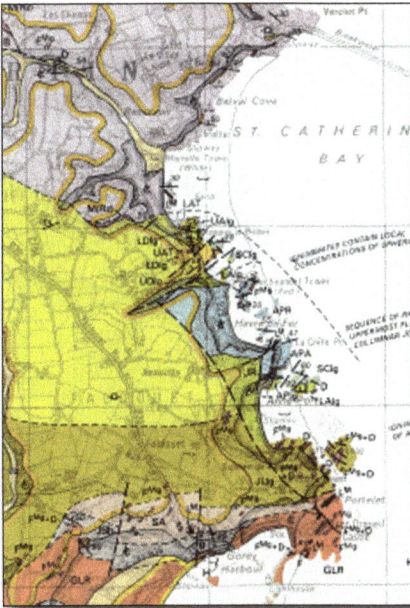

Fig. 1 (IGS 1:25,000, 1982).

The view north from the Archirondel breakwater and tower reveals the sweeping curve of La Mare Sainte Catherine (St. Catherine's Bay) (Fig. 1).

From Archirondel northwards, the seawall forming the first part of the coastline has been built over several outcrops of rhyolitic volcanic rocks both on the land and in the pebble beach for c. 400m to the small headland called Le Malade in the south central part of the bay. From here, the bay continues towards St. Catherine's slipway and tower with beach outcrops of Rozel Conglomerate.

This Trail was prompted by Arthur Mourant's papers on spherultic rhyolites (Mourant, 1932, p. 228, 232; 1933, p. 285) but there is the bonus of fine exposures of other igneous rocks and Pleistocene deposits along this part of the coast not mentioned specifically, for example, by Mourant (1933,1935) or Keen (1978) or Bishop & Bisson (1989).

Following the geological principle of studying rocks from oldest to youngest, one should start along the beach, where a variety of striking textures and structures can be seen in the volcanic rocks towards Le Malade. By ascending the sea wall, one can see a raised beach strand line of pebbles of different rock types overlain by loess and possible head deposits, resting on the eroded volcanic bedrock.

At the start of the seawall, the first outcrop is strikingly cut by white quartz veins of different thicknesses (Fig. 2), striking c. N - S and this is followed by outcrops of maroon, partly columnar - jointed rhyolites best seen from the

top of the sea wall adjacent to the start of the path; these are striking but less so than those to the south at La Crête Point (Fig. 3).

Then, descending and continuing along the beach there are striking textures and patterns to be seen in scattered outcrops of flow - banded rhyolites with

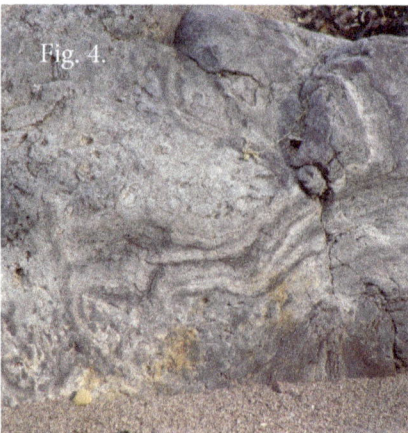

streaked ignimbrite intervals, breccia beds, tuffs and flow - folded intervals (Fig. 4), and there are scattered exposures of spherulitic rhyolite in addition to those around Arthur Mourant's sites (Fig. 5).

Outcrops of spherulitic rhyolites are rarer than at Bouley Bay and La Tête des Hougues on the north coast, and beach specimens can not always be related to

an outcrop. In between the central and northern outcrop of Le Malade, there is the bonus of a c.1m wide mica lamprophyre dyke, striking N - S, which has been eroded to form a shallow gully in the beach, the low sides of which show the contacts are both vertical and curved (Figs. 6, 7).

Then here, there are the raised beach and loess deposits. By climbing onto the seawall by the easiest route, one is confronted on the landward side, by a series of exposures, almost continuous, of a thin, raised beach strand with rounded pebbles of pink granite, black and white speckled diorite and maroon rhyolite (Figs. 8, 9).

These lie on the eroded surface of the underlying rhyolite bedrock and also contain concentrations of angular rhyolite fragments and in some places are overlain by deposits of mixed loess and head (Fig. 10). The last exposure of note in this section is a small section of uniformly yellow - brown loess above the raised beach deposit in the bank on the north side of the drainage channel through the seawall (Fig. 11). All these exposures not only reveal the uneven nature of the unconformity between the rhyolites and the superficial deposits due to differential erosion along the fracture lines of the joints and faults, but

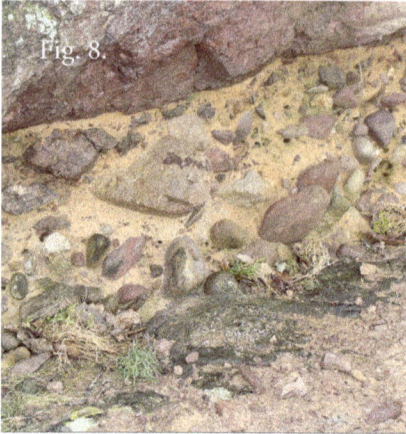
Fig. 8.

Fig. 9.

Fig. 10.

Fig. 11.

also the variety of rocks types from different source areas in Jersey and the more recent deposits below the soil horizon.

At the southern end of St. Catherine's Bay, just north of Le Malade, in the upper part of the present beach, thicker outcrops of yellow - brown silts, clays, sands and grits occur (Figs. 12, 13), possibly preserved in and on the eroded junction between the Bouley Rhyolite and the Rozel Conglomerate Formations. These too may be covered by present day beach deposits after certain tides. They are stratified and the layers vary in thickness but the depths to bedrock have not been measured, nor have stratigraphic sections been drawn.

Fig. 12.

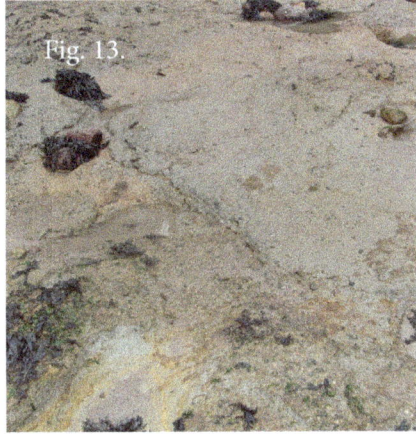
Fig. 13.

To Le Houguillon.

In the north central part of St. Catherine's Bay, Le Hougillon is approached along the coastal cliff path over outcrops of the Rozel Conglomerate from the slipway at St. Catherine's Tower (Figs. 14, 15) and also cropping out on the beach to the south. Here, excellent examples of different local rocks are exposed, especially of the shale and the rhyolites. Interestingly, none of the granite pebbles is of a local granite and they are thought to be from a granite since eroded or presently unexposed. The pebbles vary in shape, size and angularity, and represent a flash flood deposit within a desert environment after uplift and folding of Jersey's older rocks.

Fig. 14.

Fig. 15.

Fig. 16.

Fig. 17.

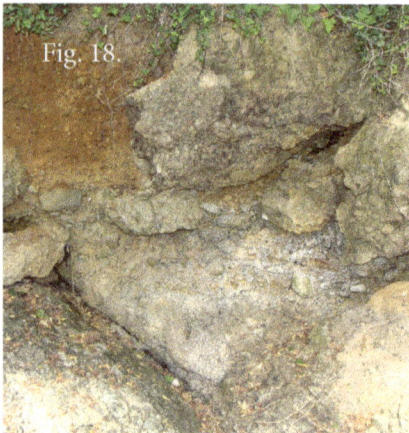

Fig. 18.

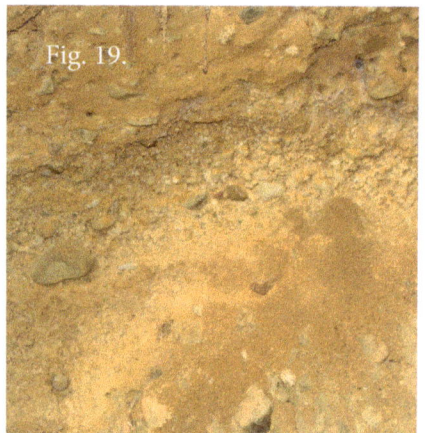

Fig. 19.

There are more 8m raised beach deposits of rounded pebbles in yellow sand and grit resting in eroded hollows and crevices of the eroded underlying Rozel Conglomerate (Figs. 16 - 19). These are presumed to be near the one mentioned at Bel Val Cove further north by Keen (1978) (Figs. 6 - 7). Overlying, the raised beach deposits are mixed head and dark grey soil containing oyster shells and scattered charcoal. These are thought to be slumped deposits from the time of the roadworks above the outcrop, when the coast road was constructed to St. Catherine's Harbour.

St. Catherine's Viviers.

This raised beach is part of a mixed section where loess, head and the beach pebble strand are exposed on the south east side of Le Verclut hill, immediately alongside the western side of the Viviers building. The raised beach (Figs. 20, 21) occurs in the lower part of the outcrop and is overlain by the thickness of head and loess (Figs. 22, 23). The distribution of angular fragments and the yellow - brown sand, silt and clay of the loess is variable in contrast to some other head deposits which are more stratified.

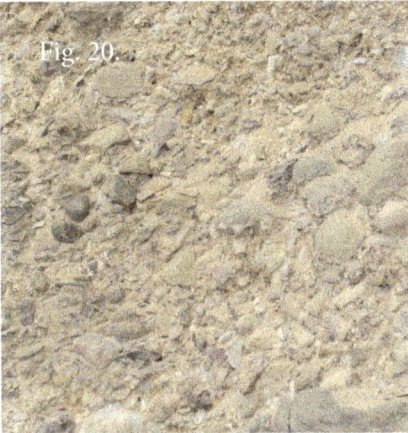

Fig. 20.

Fig. 21.

Fig. 22.

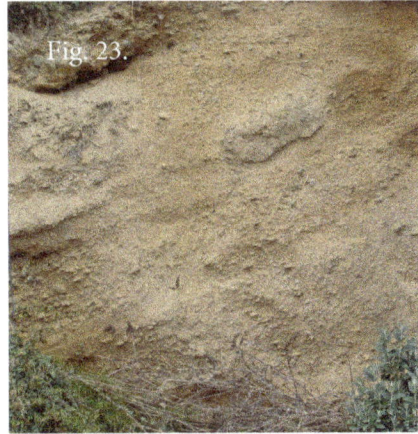

Fig. 23.

Brief Geological History

The volcanic rocks of breccias, ignimbrites and flow - banded rhyolites were formed during obvious volcanic episodes of acid volcanic activity during the Lower Palaeozoic c. 477 – 450Ma (Ordovician). The flows were then uplifted, folded and eroded during a period of mountain building and then overlain by the conglomeratic deposits of the Rozel Conglomerate to the north also during the Lower Palaeozoic, c. 450 – 427 Ma (Silurian).

The time interval between the folding and erosion of the volcanic rock formations and the deposition of the raised beach and loess deposits seems to have been enormous. If any rock units other than those of the Rozel Conglomerate were deposited, there is no trace of them, not even of any of the Eocene limestones (56 - 35 Ma) which occur nearby forming the bedrock of La Déroute channel to the east.

The raised beach deposits form part of an 8m sea level rise during the last interglacial (Ipswichian, c.130 – 115,000 years ago) while the loess and head deposits form part of the wind - blown and gelifuction deposits, formed during the last glacial of the Pleistocene (Devensian, c.115 – 10,000 years ago).

References.

Bishop, A. C. & Bisson, G. 1989. *Classical areas of British Geology: Jersey: description of 1:25,000 Channel Islands Sheet 2. (London HMSO for British Geological Survey).*

Keen, D. H. 1978. *The Pleistocene deposits of the Channel Islands. Rep. Inst. Geol. Sci. No. 78/26.*

Mourant, A. E. 1932. *The spherulitic rhyolites of Jersey. Mineral. Mag. Vol. 23. pp. 227 – 238.*

Mourant, A. E. 1933. *The geology of eastern Jersey. Q. J. Geol. Soc. London. Vol. 89. pp. 273 - 307.*

Mourant, A. E. 1933. *The raised beaches and other terraces of the Channel Islands. Geol. Mag., Vol. 70, pp. 58-66.*

Mourant, A. E. 1935. *The Pleistocene deposits of Jersey. Ann. Bull. Soc.Jersiaise. Vol. 12, pp. 489-496.*

Did you know...

That there are granites at the SE, SW & NW at the corners of the island; but there are 6 types with diorite in the SE Granite, 3 types in the SW Granite and 4 types with diorite in the NW Granite, all of different ages?

The Beauport Trail.

Granite ramparts, dykes, raised beach, loess & head.

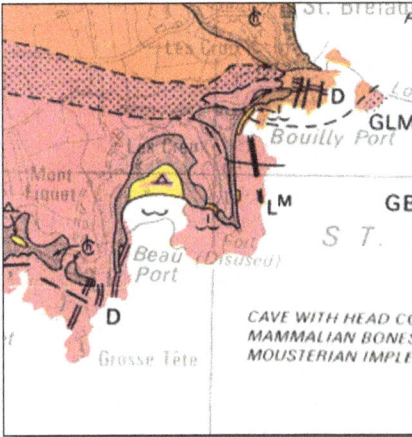

Fig. 1.

Beauport (Le Beau Port) is a small, southerly facing bay to the west of St. Brelade's Bay (Fig. 1) in the south west of Jersey.

It is a strikingly picturesque bay framed by cliffs and a varied skyline below which there are almost continuous outcrops which provide a snapshot of the changes in Jersey's geological history.

It is reached from the car park on top of the cliffs, by paths winding down the steep hill at the back of the bay, which levels out to a gentle slope covered by fern and grass on sand (Fig. 2) and which finish above low cliffs of angular granite fragments of all sizes set in yellow silt and sand. Steps then lead down the low cliffs onto a beach of very smoothly rounded, pink granite pebbles.

Fig. 2.

This soon gives way to a lovely yellow to pink, sandy beach in a truly sheltered, impressive bay, guarded by granite ramparts, and surrounded by orange to pink and red rocks, eroded

Fig. 3a.

Fig. 3b.

Fig. 3c.

into different shapes. It is backed by the stable slopes above low cliffs of yellow silt and sand, loaded with angular granite fragments of all sizes.

Not only is the scenery spectacular, with the cliffs, pinnacles, gullies, platforms, stacks and reefs, a result of continued wave erosion, but the geology is also very interesting and easily seen in a small area. Its variety extends from uniform, red to pink microgranite with rare inclusions of diorite, to red, brown and yellow granite, the colours of which are due to chemical weathering along the many joints, and to divided grey dolerite dykes, then to a section of raised beach pebbles, loess and glacial head.

The bay has been eroded in a well - jointed, uniformly crystalline microgranite (Beau Port or St. Brelade's granite), one of three varieties in the SW granite which crops out between La Corbière to Noirmont Points.

The bay is guarded by a low rocky platform below cliffs to the east, by a lone 'sentinel' rock, or stack in the centre, and by the battlement - like ramparts of La Grosse Tête cliffs to the west (Figs. 3 a, b, c).

The granite is well - jointed with three principal joint directions at right angles, but also with inclined joint planes cutting across them; these subsequently influenced weathering and erosion, being weaker parts of the rock, and thus determined the shapes

Fig. 5.

of the clefts, gullies, pinnacles and stacks (Figs. 4, 5).

The granite was later intruded by N - S striking dykes, grey walls of rock within the granite along the beach rock platform of the western cliffs (Figs. 6, 7). Interestingly, these are single examples of a period of minor intrusion, whereas others may be seen cutting across earlier E - W striking dykes to the east in Bouilly Port.

The beauty of these exposures is that they show variations in the relationship of the dyke to the granite and the variation from partial and incomplete which took place during the intrusion, so that attenuated dolerite slivers interdigitate with the granite (Fig. 8), and parts of the dyke are still under a roof of granite (Fig. 9).

On the eastern side of the bay, the trail begins with the outcrops of pink, uniformly crystalline quartz and feldspar microcrogranite, which

Fig. 6.

Fig. 7.

21

Fig. 8.

Fig. 9.

Fig. 10.

has incorporated rare xenoliths of an earlier black and white speckled diorite, stained brown by the oxidation of the iron in the hornblende, seen on the eastern raised cliff platform (Fig. 10).

Prolonged weathering and erosion have removed the surrounding and overlying country rock during millions of years (as there are now no overlying roof rocks) and we can see how the granite has been affected by exposure to the elements.

Chemical weathering effects are localised around joints. These included oxidation and hydration, two types of chemical weathering, which affected the iron in the feldspars and produced very varied and striking structures or patterns, a form of Liesegang banding not yet seen elsewhere (Figs. 11, 12).

In the cliffs at the back of this platform, is a raised pebble beach overlain by angular head and lenses

Fig. 11.

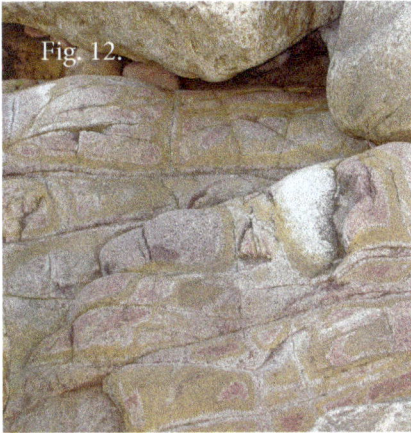
Fig. 12.

Fig. 13.

Fig. 14.

of loess. This is an example of the 8m beach which crops out in many cliff sections around the island (Figs. 13, 14). It was caused by a rise in sea level which also produced 8m wave - cut notches in the steeper cliffs. Three periods are known from their deposits, during which beaches were produced at 30m, 18m and 8m in the island, but only the 8m one is seen here.

Here, the uneven nature of the erosion surface (unconformity) below the raised beach can be seen as well as the varied brown, sandy and gritty matrix in which the pebbles are embedded. Even eroded joints in the platform have been filled with pebbles in some cases (Fig. 15).

The yellow - brown, silty loess above, was deposited during falls in sea level (regression) which produced dry, wind - dominated, terrestrial landscapes when the bedrock broke into angular fragments under freeze -

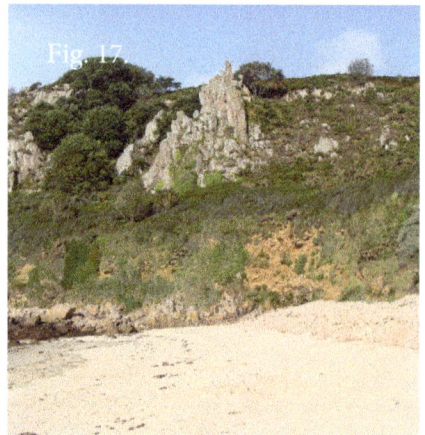

thaw conditions. These were deposited to form glacial head during gelifluction (Fig. 16).

The sequence of head seen at the back of the bay continues on the western side (Fig. 17) but does not reveal any raised beach deposits. This may be because the deposits were unevenly deposited or because subsequent erosion removed them...or both may have happened. The greatest thickness of head is recorded as c.15m here (Bishop & Bisson, 1990, p. 87, 89).

Brief Geological History

The granite is the second of three granites to have formed from the intrusion of the SW granite mass into the Jersey Shale Formation, its small uniform crystals showing that it cooled quickly.

No evidence of the subsequent SW to W striking dolerite dyke swarm, seen nearby, has been found, but the Beau Port granite was intruded later by other N - S striking dolerite dykes which were emplaced unevenly along joint planes.

Later again, prolonged weathering and erosion removed the surrounding and overlying country rock during millions of years, as no overlying roof rocks can be seen.

Later, more chemical weathering occurred, especially during the more tropical climatic regimes when the Cretaceous (Mesozoic) seas caused the deposition of chalk to the north and west of us, and later again, when the

Eocene (Cenozoic) seas caused the deposition of fossiliferous limestone partly surrounding us.

This weathering included oxidation and hydration, two types of chemical weathering which affected the iron in the feldspars and produced the varied and strikingly coloured patterns.

Periods of sea level rise and fall (marine transgression and regression) then followed during the Ice Ages of the last two million years, until c. 12,000 years ago (c. 10,000 BC). The withdrawal produced a dry, wind - dominated, terrestrial landscape when yellow - brown, silty loess was deposited and the bedrock broke into angular fragments, under freeze - thaw conditions, which were deposited to form glacial head during gelifluction. Erosion by wave action during the last 5,000 years when the low flat - topped hills became the islands they are today, has accentuated the bay and wave refraction round the headlands has caused longshore drift of sand alongside the cliffs which was deposited to produce the beautiful beach. Accompanying wind action transported the blown sand seen along the pathways (Bishop & Bisson, op. cit. p. 93).

References.

Bishop, A. C. & Bisson, G. 1989. Classical areas of British geology. Jersey. Description of the 1:250 000 Channel Islands Sheet 2. British Geological Survey. HM Stationery Office, London.

Did you know...

That the Geological Cycle describes the first rocks as Igneous, followed by weathering and erosion to form Sedimentary rocks, then by heat and pressure to form Metamorphic ones?

La Bonne Nuit Trail.

La Bonne Nuit Raised Beaches Revealed.

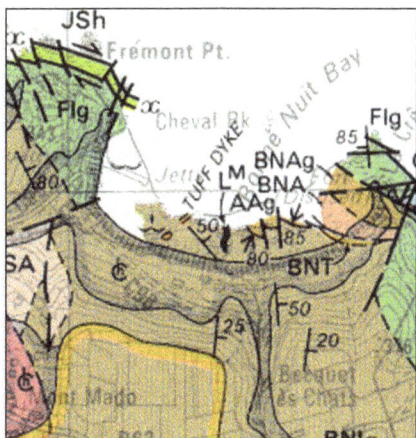

Fig. 1.

At Bonne Nuit, the raised beach deposits are exposed in the cliffs above the beach rocks of the bay (Fig. 1.) on the north coast of Jersey in the parish of Trinity. The bedrock was examined in 2005 as part of a study of Dr. Mourant's work on the spherulitic rhyolites. They were found to be ignimbrites with possible spherulites and flattened pumice textures (Figs. 2, 3).

They form part of the St. John's Rhyolite Formation with the spherultic rhyolites occurring further east.

The raised beach deposits in the cliff section at the back of Bonne Nuit Bay (Figs. 4, 5) are a striking example of the effects of sea level rise and fall during the later stages of the last ice age, and rank with other striking ones seen by the author at Portelet Bay and on the west side of Noirmont south of La Cotte de St. Brelade.

The cliff section however, which crops out above the ignimbrites at the back

Fig. 2.

Fig. 3.

Fig. 4.

Fig. 5.

of the bay was not examined until later visits with J. Sonnex during her UCL Birkbeck Honours dissertation studies in 2009 and with Dr. Martin Bates (UW Lampeter) during his Quaternary geoarchaeology studies of La Cotte de St. Brelade in 2010. The cliff section is very well exposed and reveals a thick sequence of Pleistocene deposits with a fine cross section and stratigraphic sequence of raised beach, loess and head facies.

From the slipway at the western end of the bay by the jetty, and from the eastern end of the beach wall it was found that within the beach and the cliff, there are three raised beach pebble beds exposed representing former sea levels, two in the present day beach and one in the cliffs at the back of the beach.

The first or lowest exposure is of a ferruginous, pebble and coarse sand deposit at c. mid - tide level. The second, next highest, exposures are of rounded pebbles in a brown to green, silty clay matrix (not sampled) almost hidden below the boulders approximately half way up the present beach.

The third, is higher again, and occurs in the lowest part of a sequence of various yellow - brown silty clays, possible loess, and head, and seems to be a strand of pebbles which are matrix-supported, near the base of the cliffs at the back of the beach; this is possibly the 8m raised beach.

The first raised beach deposit crops out in the lower part of the present day Bonne Nuit beach and lies on bedrock and underneath the present large boulders.

It is a brown, well - cemented ferruginous pebble deposit with laminated coarse sand layers which lies on and against the bedrock c.10 - 15m from the end of the sea wall and c. 40m down beach (Figs. 6, 7).

In December 2012, a striking view of the deposit was obtained as present day backwash had removed the beach shingle and a more extensive outcrop was revealed which showed a thicker deposit, c.1m of the conglomerate lapping around prominent parts of the uneven, eroded surface of the rhyolite bed rock.

The deposit is unsorted and the pebbles vary in size. They are well - rounded but of low sphericity and are grain - on - grain as well as matrix supported. They consist of a mixture of local rocks, eg. the underlying ignimbrite but also of grey shale or dolerite pebbles (not sampled) and some larger, c. 10-15 cm rounded quartz pebbles. The laminated (bedded) coarse sandstone and fine grit parts of the sequence form lenses of varying thickness, shape and extent between the poorly sorted pebble intervals.

This deposit was the lowest deposit found and seems to occur lower down the beach (no levelling was done) below the second type of which no previous description has yet been found but it may not have been exposed in 1978 (Keen, 1978).

This well - cemented ferruginous conglomerate is assumed to be the one described by Mourant (1933, p. 59) as "a hard ferruginous conglomerate"

Fig. 6

Fig. 7

Fig. 8

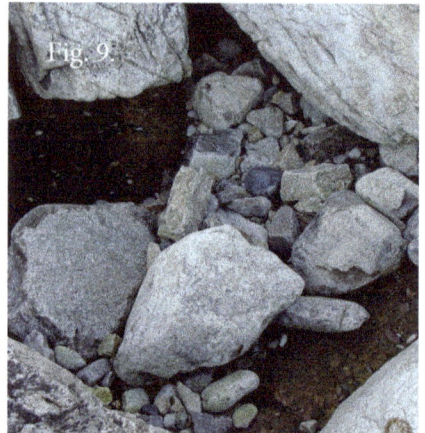

Fig. 9

and "a cemented ancient beach deposit" (Mourant, 1935, p. 489). It is also described by Keen (1978, p. 9) as the only beach gravel to be seen and is " a coarse, iron - cemented, sandy shingle....attached to the rock platform" 45m seaward of the modern cliffs. Mourant (1935, op. cit.) thought that the deposit "may belong of the 25ft. raised beach" but that it is possible it may "represent a distinct horizon". It was also described by Bishop & Bisson (1989, p. 85 - 86) in the 8m beach descriptions as 'beach gravel strongly cemented by ferruginous minerals'

The second raised beach deposit is thought to form a separate deposit because the outcrops are exposed in several places slightly higher up the beach between 10 - 20m further east (again no levelling was done), that is, part way between the ferruginous well - cemented deposit and the cliff deposits.

These deposits are similar to each other and are at the same elevation. They are exposed in cavities under large boulders as if having been eroded away as the boulders were undercut by wave action. In contrast to the red - brown ferruginous conglomerate, they are non - ferruginous and semi - indurated pebble deposits seemingly in a dark grey - brown soil - like, sandy clay groundmass (not sampled due to difficult access). They are often green - coloured due to a thin covering film of algae (Figs. 8 & 9). Given the seemingly different elevations between the two different beach gravel/pebble conglomerates and their position below of the strand line of pebbles in the superficial deposits of the cliff, are they each separate deposits, or are they parts of the same 8m beach deposit which have formed under differential physico-chemical weathering and lithification conditions due to either their different positions in the beach, or to those different conditions prevailing at different times?

The third raised beach deposit (8m) crops out in the cliff sequence as a well - exposed, excellent outcrop of a thin strand line of rounded pebbles just above the base, with interbedded yellow, silty clays (some laminated), other thin strands of raised beach pebbles and gravel and glacial head layers (Fig. 10).

The section exhibits well - sorted intervals alternating with intervals containing scattered large angular blocks all along the back of Bonne Nuit Bay. The pebbles are well rounded but of variable sphericity, some being oval, and are poorly sorted, that is, they are matrix-supported being surrounded by brown - yellow clay - silt, and rarely show grain - on - grain structure (Fig. 11). The strand-like appearance is not uniform along the

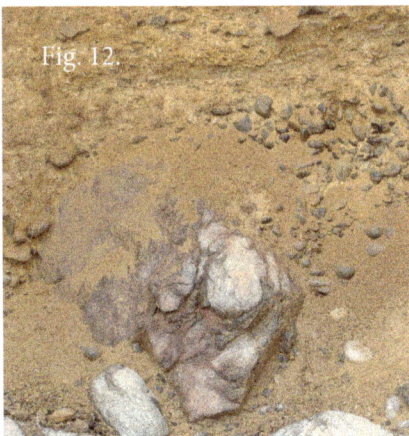

Fig. 14.

Fig. 15.

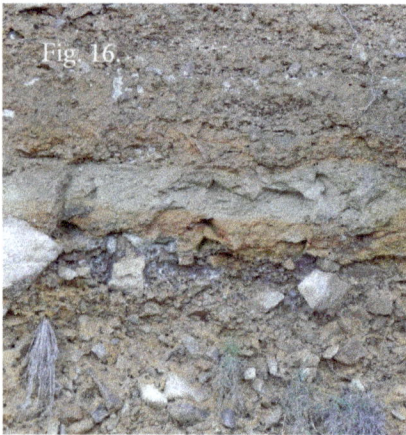
Fig. 16.

outcrop and variation in the form of the raised beach deposits occurs at the eastern end of the cliff where one can descend from the cliff top beyond Le Cheval Roc via a steep path to the eastern end of the beach. For example, there are gully accumulations (Fig. 12) of pebbles and silt lenses from a possible lagoon/pool (Fig. 13) which lie below pebble deposits & head along this part of the outcrop.

Above the raised beach and possible lagoonal deposits, beds of glacial head occur between a dipping, lower laminated silt (base) and a grey silt bed which cuts across it; some of these intervals show wedging or thinning (Fig. 14) where a mid-section, pale grey bed, c. 30cm thick provides a good marker half way along the bay. Some layers seem to be truncated by overlying beds (Fig. 15).

The whole sequence seems to represent periods of marine advance (pebble strand); lagoon-pool conditions; fluvial deposition, for example, the apparent cross-bedded part; followed by later glacial head formation (gelifluction deposit) and some thin loess (aeolian) deposits. The raised beach deposits have also been described by Bishop & Bisson (1989, p. 83) who also interpreted the head deposits as forming under variable conditions (op. cit. pp. 85 - 88).

The type of iron is not named, nor have the conditions of formation been discussed, that is, is it an example of subaerial or shallow depth diagenetic ferruginisation, or of leaching and precipitation of an Fe-rich leachate by ground water as in the present cliff (Fig. 16), but within a since-eroded superficial deposit? The intermediate level deposit is much less indurated as are the pebble deposits at the base of the cliffs.

Summary

The possibility that the three pebble deposits represent three former sea levels and three distinct beaches was raised by Arthur Mourant when he stated that the ferruginous deposit which crops out "at a little above mean tide level... may belong to the 25ft. raised beach but it is possible that they represent a distinct horizon" (Mourant, 1935, p. 489).

This raises the question that given their elevations above the Fe deposit, are the other pebble deposits also distinct horizons, therefore representing pauses in sea level rise below 25ft.(8m), separated by periods of deposition under sub - aerial conditions. The well - cemented ferruginous deposit seems to be much older, with time allowed for possible oxidation and lithification to occur.

The type of iron is not named, nor have the conditions of formation been discussed, that is, is it an example of subaerial or shallow depth diagenetic ferruginisation, or of leaching and precipitation of an Fe - rich leachate by ground water as in the present cliff, but within a since - eroded superficial deposit? The intermediate level deposit is much less indurated as are the pebble deposits at the base of the cliffs.

Could they represent parts of a diachronous deposit with three periods of still - stand to permit beach formation, and therefore the lower two are evidence of earlier raised beaches than the 8m one?

Brief Geological History

In this part of Jersey, after the volcanic episodes which produced the rhyolites and ignimbrites (St. John's Rhyolite Formation) and the overlying spherulitic, flow - banded rhyolites (Bouley Rhyolite Formation), the intrusion of the NW Granite (or Igneous Complex) occurred and was followed by uplift and erosion to produce the Rozel Conglomerate Formation.

There then seems to have been a long period of weathering and erosion, deposition of Tertiary Eocene limestones around us and then several regressions and transgressions during the Pleistocene glacial and interglacial periods which produced interbedded loess, head and raised beach deposits respectively.

The raised beach deposits of Bonne Nuit Bay are varied and seem to represent a variety of conditions, not always seen in other deposits around the island. The variations in the deposits south of La Cotte de St. Brelade seem to bear comparison with the ones described above starting with a lower, well - cemented former beach shingle. The sequence of raised beach, loess and head is thought to represent the change from interglacial to glacial conditions with the advance and retreat of the sea, but also may have been combined with the rise and fall of the land due to partial isostatic rebound.

The ferruginous cement of the lowest mid - tide beach could have formed from iron transported in solution by sea water or groundwater, the nearest source of iron being the underlying volcanic bedrock. It was then possibly oxidised under subaerial conditions to form the present very well - cemented ferruginous beach deposit seemingly before the deposition of the upper raised beaches.

References.

Bishop, A. C. & Bisson, G. 1989. Classical areas of British geology; Jersey: description of 1:250,000 Channel Islands Sheet 2. London HMSO for British Geological Survey..

Brown, M. Power, G. M. Topley, C. G. & R. S. D'Lemos, R. S. 1990. Cadomian magmatism in the North Armorican Massif. p. 181 - 213. In The Cadomian Orogeny. Eds. D'Lemos, R. S., Strachan, R. A. & Topley, C. G., 1990, Geological Society Special Publication No. 51. Geological Society, London.

Keen, D. H. 1978. The Pleistocene deposits of the Channel Islands. Rep. Inst.. Geol. Sci., No. 78/26.

Lees, G. J. 1990. p. 273 - 291. In The Cadomian extensional magmatism in Jersey, Channel Islands, p. 273 - 291 in The Cadomian Orogeny, Geol. Soc. Spec. Publ. No. 51. (see D'Lemos et al. above).

Marett, R. R. 1911. Pleistocene Man in Jersey. Archaeologia, lxii, pp. 449-480.

Mourant, A. E. 1933. The Raised Beaches and Other Terraces of the Channel Islands. Geol. Mag. Vol. LXX, pp. 58-66.

Mourant, A.E. 1935. The Pleistocene deposits of Jersey. Bull.of the Soc. Jers.Vol.XII, pp. 489-496.

Renouf, J. 1986. Geological setting and origin of La Cotte de St. Brelade, p. 35-52, in (Eds.) P. Callow & J. Cornford. La Cotte de St. Brelade: 1961-1978, Excavations by C.B.M. McBurney.

Renouf, J., James, L. 2010. High level shore features of Jersey (Channel Islands) and adjacent areas. Quaternary International (2010), doi: 10.1016/j.quaint.2010.07.005

Le Bouilli Port – Les Creux Trail

Raised beach, dykes & chasms (Les Creux Fantômes).

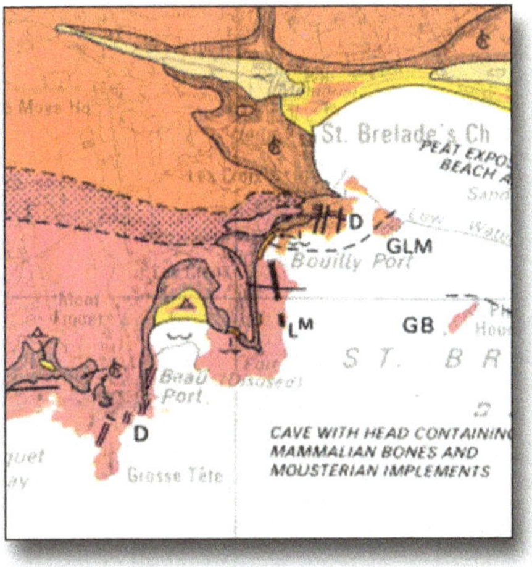

Fig. 1.

This is a beautiful little area which was explored by the author prior to a Section visit; it is scenically attractive and spectacular in parts, and also historical. Although named on the OS map, as La Saline, it is situated on the western side of St. Brelade's Bay between St. Brelade's Church and Les Creux on land, and La Saline and the Slipway on the coast (Fig.1).

It includes outcrops along Le Chemin de(s) Creux beyond St. Brelade's Church and La Chapelle des Pêcheurs, southwards into the rocky bay of La Saline, Le Bouilli (Stevens et al, 1985) or Bouilly Port (Perry's Guide), and the beach outcrops from Bouilly Port northwards to the Slipway.

The road outcrops along Le Chemin de(s) Creux are easy to examine but to examine the beach outcrops, return to the church, descend to the beach and use the concrete path to the jetty. Climb the steps to the Viewpoint (site of Le Coleron Battery, again unmarked on the OS map) above Bouilly Port (Le Bouilli means 'on the boil', a reference to the wave action), and to your immediate right (west), is a spectacular, deep, steep-sided chasms, which clearly reveals some of the geology around you.

To examine the spectacular chasms mentioned above, descend to the bottom of the steps, turn south and cross the wave-cut platform to round a low rocky point into Les Creux Fantômes (phantom hollows), named after the weird

Fig. 2.

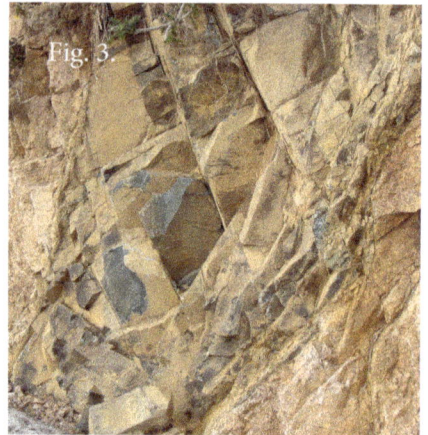

Fig. 3.

shapes of the rocks (Stevens et al, 1985) at the northern end of Bouilly Port. Care should be taken on each of these routes over uneven surfaces and should only be done at low spring tides.

This trail can start along Le Chemin de(s) Creux, the private road behind St. Brelade's Church to Les Creux, recently widened a little, with fresh cliff - face exposures on one side, and views back to the view-point and down into Les Creux Fantômes (Bouilly Port) on the other. Geologically, the area is situated in the Southwest granite (Bishop & Bisson, 1989), actually three different granites which crop out between La Corbière and Noirmont Points, and the road passes near where two of these granites are exposed. The area also includes several dolerite dykes, some clearly exposed in the roadside cliff face, and others in the chasms, also a new raised beach exposure and glacial head deposits, a variety of erosional features and at certain times, beach springs.

Starting along Le Chemin de(s) Creux past the church, the pink, coarsely crystalline granite, often porphyritic, crops out in freshly exposed faces. Well - jointed sections are followed by two minor intrusions of dark grey dolerite (Figs. 2 & 3). The first is low down in the face and has two thin 1cm thick veins rising at a low angle towards the second which is a 1.5m wide dark grey, dolerite dyke, striking c. E - W and steeply dipping southwards.

Further along the road, there is an unusual exposure of what appears to be a fault zone possibly formerly eroded into a gully, c.1.5m wide with angular fragments of granite but also with areas of rounded beach pebbles near the base left and right of a large central, rounded boulder (Figs. 4 & 5).

This appears to be a new raised beach site at c. 30m level (J. Renouf, pers.

Fig. 4.

Fig. 5.

Fig. 6.

Fig. 7.

comm.). The pale brown – pink, rounded granite pebbles are not abundant and vary in size from 1 - 10cm, lying within a coarse sand and grit matrix; hence they form an unsorted deposit. Overlying it there is a deposit of mixed angular granite fragments lying within a grit and sand matrix, bearing similarities to glacial head. The deposit is bounded on each side by vertical walls of granite, 1 – 1.5m apart and appears to fill therefore, a narrow gully, possibly in a former wave - cut platform which lies above and is covered by the present day soil and vegetation.

It is difficult to descend to the beach from this road (as mentioned above); the path is not marked and descends through trees. It is steep, twisting and stony in parts and involves a scramble down over rocks at the base onto the rocky beach. The safer route is via a narrow path which descends past the last house. This leads onto a short path to Le Coleron Battery viewpoint and

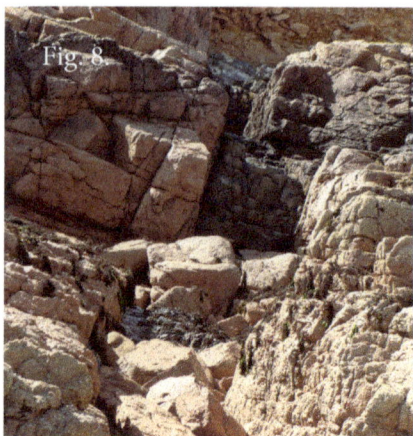
Fig. 8.

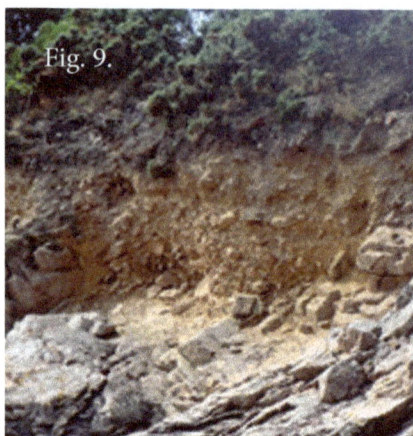
Fig. 9.

passes a spectacular deep, vertical sided, narrow chasm, eroded into the cliffs from the south (Fig. 6) striking N - S.

Just before the lookout, the path allows an ideal view immediately west into the continuation of the vertical - sided chasm, and right opposite you is a spectacularly exposed, near -v ertical dyke in the far wall (Fig. 7) striking E - W.

Being very careful, one can walk out onto part of a narrow ridge which separates the look-out chasm from another just to the west. The dyke continues below you across this ridge and across the next chasm to the west, though it is not exposed as clearly in its western wall. Looking back at the wall which descends from the viewpoint path, there is no dyke exposed along strike, so it would seem that this gorge has been eroded along a tear fault. Further exploration is needed to determine the direction of movement.

Walking back along the viewpoint path and descending to the breakwater leads onto the second part of the trail at the southern end of St. Brelade's Bay beach. Looking south one can see the present wave - cut rocky platform and one needs to walk over it to find a dolerite dyke striking across it and into the cliff face (Fig. 8). It is faulted by small N - S striking dextral tear faults and could be the missing continuation of the dyke seen from the lookout, displaced to the south by a N - S dextral tear fault along which the chasm has been eroded. The cliff itself is composed of well - jointed pink granite with an irregular upper surface below a variable thickness of glacial head, part of which shows weak stratification (Fig. 9).

Scrambling up and over the low cliffs (Les Fantômes) and little headlands

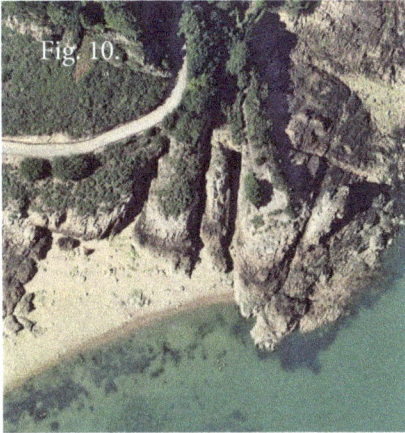

Fig. 10.

Fig. 11.

Fig. 12.

further south, one descends into the area of Les Creux Fantômes and is confronted by spectacular chasms c. 2m wide, with near - vertical walls (Fig. 10), eroded in the granite and striking N - S. There are three, the first one having been seen alongside Le Coleron viewpoint path.

Remnants of dark grey, dolerite dykes crop out in the floors of them and also in the walls. They vary a little in thickness (Figs. 11 & 12) and their intrusive contacts with the granite are clearly exposed. The contacts vary from planar to slightly curved and there are thin offshoots or bifurcations which vary in length. Unusually here, there is also an example of veining in the floor of the eastern gully, where an almost filigree structure, clearly showing the mode of intrusion, occurs in an apparent boulder or possibly a remnant of a dyke buried by the pebble deposit (Fig. 13).

From here, walk back over the wave - cut platform to Le Coleron breakwater and along the concrete path to just beyond the steps down the low breakwater. Here, depending on the work of tides and waves, a study of the beach gravels in the first few metres, will reveal yellow - brown, silty clay around the base

Fig. 13.

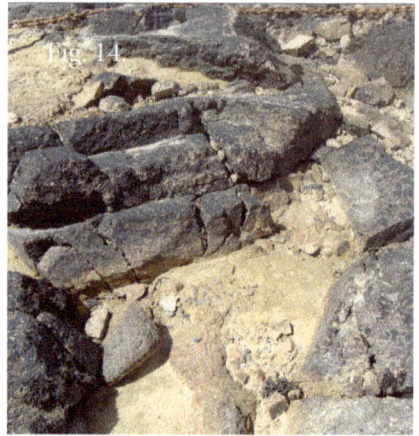

Fig. 14

of small granite outcrops and forming patches within the recent gravels. This was more extensive during the storms in March ('08) but seems to represent a buried loess deposit (Figs. 14 & 15).

It is doubly interesting because a variety of differently coloured, worked flints lie on the surface and rarely, only partly in the deposit. In addition, isolated pieces of possible Normandy ware can be seen and a small part of the stem of a clay pipe was also found.

Finally, a walk across the beach towards the slipway reveals small streams which seem to come from the culvert to the left of the slip. However, some originate in the beach a little further to the right (east) and are worth examining. These streams come from water bubbling up through the sand

Fig. 15.

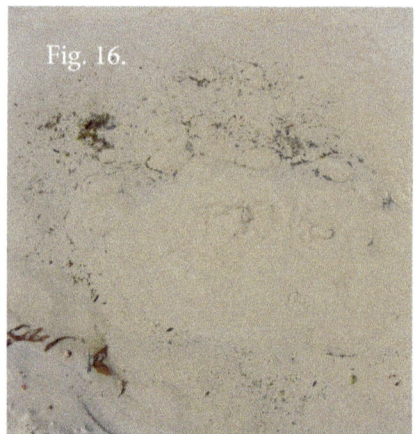

Fig. 16.

and seem to represent small springs of ground water from the granite bedrock below the beach and forming the land to the north. They also have brown deposits on the surface, possibly iron rich bacterial blooms (Figs. 16).

Brief Geological History.

During the Lower Palaeozoic, the Porphyritic granite (c. 550 Ma) and the Aplite/Microgranite (c. 527 Ma) of the SW granite in the area described, were intruded below the Precambrian Jersey Shale Formation after its deposition, uplift and folding, and during the eruption of the andesites forming the lower part of the Volcanic series (c. 522 – 477 Ma). The adjacent Corbière granite was then intruded (c. 483Ma) and more andesites erupted, while during the same period (but later than 483Ma) there was intrusion of the dolerite dykes of the Main Dyke Swarm along E - W strikes.

The N - S dolerite dykes were intruded later possibly at the same time as the N - S tear-faulting as they cut the E - W dykes.

Intrusion of the NW granite, occurred below, during the eruption of the rhyolites and the uplift, erosion and deposition of the Rozel Conglomerate (c. 477 – 426 Ma).

The folding and intrusion occurred during a period from c. 700 – 425 Ma, ie. c. 275 Ma from the Precambrian to Silurian, a long period known as the Cadomian Orogeny (Brown et al, 1990, p. 181 et seq.).

This seems to have been followed by a long period of erosion, during the Upper Palaeozoic and Mesozoic, which removed the country rock and revealed the granites, until Tertiary limestones were deposited around the island.

During the Pleistocene there were several periods of higher sea level during interglacial times when raised beach were formed, the 30m one described above being the oldest; other raised beaches deposits occur at 8m and 18m on the opposite side of St. Brelade's Bay. These were interspersed with periods of loess and head deposition during intervening glacial times, the deposits in the present littoral zone described above being produced during the last glacial period before the present sea level rise. The present climatic regimes and weather, which control the present weathering, marine and fluvial erosion and deposition, have produced the springs and the beach sands illustrated above.

References.

Bishop, A. C. & Bisson, G. 1989. Classical areas of British geology; Jersey: description of 1:250,000 Channel Islands Sheet 2. London HMSO for British Geological Survey.

Brown, M. Power, G. M. Topley, C. G. & R. S. D'Lemos, R. S. 1990. Cadomian magmatism in the North Armorican Massif. p. 181 - 213. In The Cadomian Orogeny. Eds. D'Lemos, R. S., Strachan, R. A. & Topley, C. G., 1990, Geological Society Special Publication No. 51. Geological Society, London.

Keen, D. H. 1978. The Pleistocene deposits of the Channel Islands. Rep. Inst.. Geol.Sci., No. 78/2.

Lees, G. J. 1990. The geochemical character of late Cadomian extensional magmatism in Jersey, Channel Islands, p. 273 - 291. In The Cadomian Orogeny, Geol. Soc. Spec. Publ. No. 51. (see D'Lemos et al. above).

Marett, R. R. 1911. Pleistocene Man in Jersey. Archaeologia, lxii, pp. 449-480.

Mourant, A. E. 1933. The Raised Beaches and Other Terraces of the Channel Islands. Geol. Mag. Vol. LXX, pp. 58-66.

Renouf, J. 1986. Geological setting and origin of La Cotte de St. Brelade, p. 35-52. In (Eds.) P. Callow & J. Cornford. La Cotte de St. Brelade: 1961-1978, Excavations by C.B.M. McBurney.

Renouf, J., James, L. High level shore features of Jersey (Channel Islands) and adjacent areas. Quaternary International (2010), doi: 10.1016/j.quaint.2010.07.005

Stevens, C., Arthur, J., & Stevens, J. 1985. Jersey Place Names, Vols. I & II. Société Jersiaise.

Le Côtil Point Trail

Granites, andesite, ignimbrite, wave-cut platform, gullies & caves.

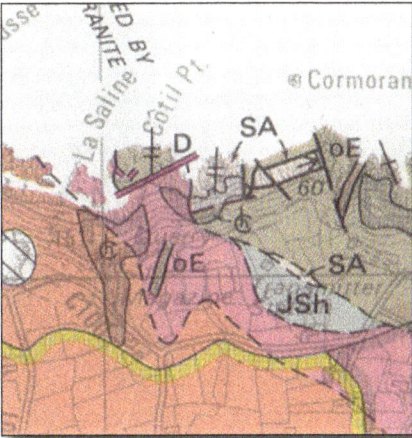

Fig. 1.

Le Côtil Point is situated on St. John's coastline just east of La Saline between La Trousse and Wolf Caves NW of the Frémont Transmitting Station (Fig. 1).

This may be considered a slightly dangerous trail so great care should be taken to wear good strong boots with appropriate soles for a good grip.

It lies below La Saline quarry and is approached through bushes and stunted trees down a very steep fisherman's path, generally scrambling and stepping sideways, but also via a ladder and a rope for two short c. 5m pitches.

The end of the path has steps and one walks onto the low rocks of the Point on the left hand side of the air photo (Fig. 2), and then on to the other rocks, heading east. These may be slippery after high tide and the green gully is generally easier for progress. The best time for access is to start c. 1 hour before MLWS to explore the outcrops in the gullies and inlets, leaving the area c. 1 hour after low tide.

Fig. 2.

The aerial photograph (Fig. 2), shows the position of Le Côtil Point on the western edge, a wide dividing inlet, and the reef (small island), off the adjacent long narrow inlets to the east, eroded into the flat wave- cut platform. The ENE - WSW striking green - floored gully in the Point and between the platform and the reef/island has been eroded along a dyke.

It was chosen because 10 references to Le Côtil Point in the BGS Report (Bishop & Bisson, 1989) show that there is a very varied geology.

A bonus for our Pleistocene geology, is a superb wave-cut platform but with several long, deep inlets eroded into it.

In addition, other geologists, Mourant and Oliver (1953) and (1958), have reported the site to contain garnets.

The geology references are as follows;

p. 14. There is an intrusive contact between the volcanic rocks Bonne Nuit Ignimbrite, St. John's Rhyolite Formation and the NW granite at OS631563. The thermal aureole is c. 300m wide with N - S vertical foliation, granoblastic textures and porphyroblastic biotite & metasomatic andradite (garnet), diopside & epidote in veins & patches, and volcanic rocks close to the granite are metasomatically enriched in K.

p. 20. Andesite (St. Saviour's) intruded by granite (see p. 52) and is overlain by ignimbrite (Bonne Nuit Ignimbrite, St. John's Rhyolite.). Just NE of former Mont Mado quarry, outcrops of andesite are terminated to the N by a pre-granite E - W sinistral (LHS) wrench. These andesites are thermally metamorphosed, and have pink alkali feldspars rather than plagioclase but no megascopic foliation.

p. 24 St. John's Rhyolite. crops out from Le Côtil to La Crête (St. John) and is a repeat of the Bonne Nuit Bay section displaced westwards by the Frémont LHS wrench.

p. 25. The Bonne Nuit Ignimbrite is 550 - 900m thick between Le Côtil & Frémont Points via Wolf Caves.

p. 26. Between Le Côtil & Frémont, the Bonne Nuit Ignimbrite is overlain by the Frémont Ignimbrite.

p. 52. NW granite has intrusive contacts with rhyolite (Bonne Nuit Ignimbrite, St. John's Rhyolite,) & andesite (St. Saviour's Andesite) at the

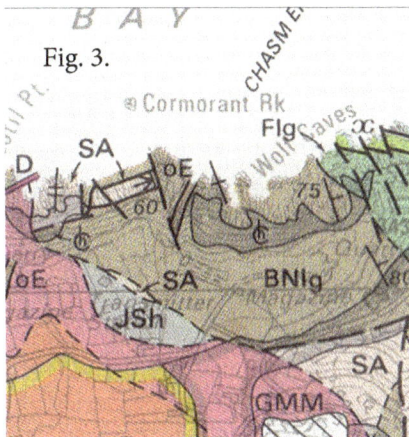

Fig. 3.

eastern most end of (the granite) outcrop in the vicinity of Le Côtil Point. (OS631562).

p. 53. At Le Côtil Point sharp bounded veins of aplogranite have intruded the ignimbrite (Bonne Nuit Ignimbrite) & the basic dykes in it. The granite (NW granite) & ignimbrite (Bonne Nuit Ignimbrite) are so similar that there is little evidence of thermal metamorphism, but the effects of K metasomatism are seen in the coatings of muscovite on joint surfaces in the ignimbrite.

p. 54. Small quarry (La Saline?) above Le Côtil Point OS6305 5602, shows contact between Mont Mado aplogranite. and the coarse St. Mary's granite with c. 1m of mafic rich granite. Old Mont Mado quarry at OS637556.

p. 68. Basic dykes dolerite at Le Côtil Point (631 562) predate the Mont Mado aplogranite giving an age of c. 480Ma as part of the NW granite.

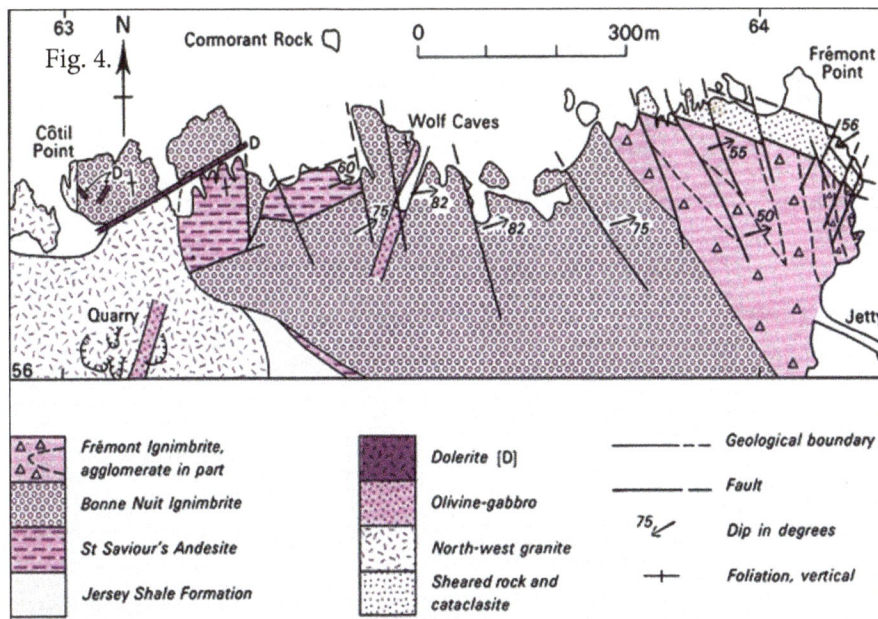

Fig. 4.

Frémont Ignimbrite, agglomerate in part	Dolerite [D]	Geological boundary
Bonne Nuit Ignimbrite	Olivine-gabbro	Fault
St Saviour's Andesite	North-west granite	75 Dip in degrees
Jersey Shale Formation	Sheared rock and cataclasite	Foliation, vertical

Fig. 3. IGS 1:25,000 Geology map (1982). Fig. 4. BGS Report map (1989) (p. 24).

p. 69. ...a dyke of olivine gabbro 8m wide occurs at Wolf Caves (6347 5616) and is presumed to continue S of Le Côtil Point.

The IGS Jersey, 1:25,000 Geology map (1982) (Fig. 3) shows the complicated geology and it is compared with the map in the BGS 1989 Report (p. 24) (Fig. 4) from which the page references are taken.

Key. GSM St. Mary's gr. GMM. Mont Mado aplogr. BNIg. B. Nuit. Ignim. SA. St. Sav. Andesite.

The following points should be noted;

1. The Mont Mado type aplogranite is not identified on BGS Report map (p. 24) (Fig. 4. above).

2. The Bonne Nuit Ignimbrite crops out on Côtil Pt. and a small reef/island (looking like a 2nd Point east) on BGS Report map; it is cut by a wide inlet dividing Le Côtil Pt. from the reef on IGS Report map, but this is not shown on the IGS 1:25,000 Geology map (Fig. 3 above) thus making the outcrop look continuous.

3. The Bonne Nuit Ignimbrite (in the St. John's Rhyolite Formation) on Côtil Point, the 1st Point.

4. The NW granite (Mont Mado type) on the W of the Point, and south of it, separated by a dolerite dyke.

5. The St. Saviour's Andesite, E of the Point in 1st inlet (in contact with the NW granite (Mont Mado type), south of the dolerite. dyke).

Fig. 5.

Fig. 6.

Fig. 7.

Fig. 8.

6. A little further east at OS6318 5620, veins of the granite (NW) have intruded and bisected the St. Saviour's. Andesite, and garnets (Andradite, Ca Fe garnet) of hydrothermal origin have been recorded in this area (Oliver, BSJ 1958, p. 181 - 183). Oliver refers to previous reports (Mourant, BSJ 1938, BSJ 1953).

The trail starts by taking great care descending the path!

The first view during the descent, is a striking one not previously described, of an impressive wave - cut platform backed by cliffs to the east (Fig. 5) and a pebble and boulder - filled bay backed by cliffs. This is the 8m platform and seems to be confirmed by the perched stack RHS (Fig. 6). A cave high in the cliffs at c. 8m at the southern end of the first inlet may also be at the former sea level.

The Mont Mado aplogranite is recognised by its small, uniform

quartz and feldspar crystal size and muscovite mica also occurs along its joint surfaces (Figs. 7, 8).

At the bottom of the steps onto the rocks, the exposures are of the NW granite - Mont Mado type. It is well - jointed and it is on these joint faces that thin films of muscovite mica can be seen.

In traversing east, earlier erosion in

Fig. 9.

Fig. 10.

Fig. 11.

Fig. 12.

the first gully has formed a perched cave which varies from lens - shaped to generally circular inside, at about the 8m sea level in the cliffs at the southern end of the gully (Figs. 9, 10). This may have been formed at the same time as the wave - cut platform further east.

The Bonne Nuit Ignimbrite, of the St. John's Rhyolite Formation, crops out on the Point and is grey to light brown in colour but exhibits the 'banded 'streaky bacon' texture, possibly of a reomorphic type, flow, with incipient spherulite formation; it also has a paler brown to pink, banded and partly 'flocculated' texture (Fig. 11).

The St. Saviour's Andesite in contrast is medium grey with small rectangular pink to red feldspars which give it a porphyritic texture. There are several varieties of this texture and they are quite unlike the white porphyritic andesite further south around St. Helier (Figs. 12 - 14).

The junction between the granite and the St. Saviour's Andesite can be seen at the back of the second inlet with granite on the RHS and andesite on the LHS with veins and attenuated sills/dykes intruding it (Figs. 15, 16).

The andesite is also intruded by a rare thick aplite sill and by several dykes and veins some with anastomosing structures, seen both in the joint and fault planes forming the sides of the inlets and the base of the platform surface (Figs. 17 – 20).

Fig. 13.

Fig. 14.

Fig. 15.

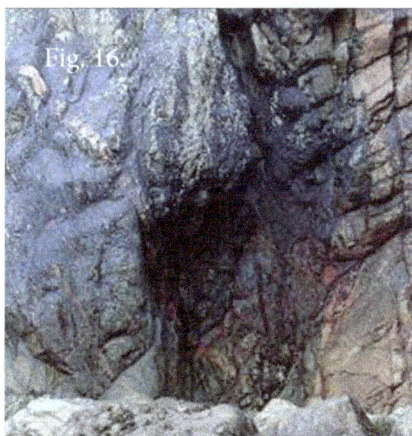
Fig. 16.

And then there are the elusive garnets described by Mourant and Oliver as follows;

Mourant (1938, BSJ. p. 289) recorded veins of 'massive brown garnet' with 'occasional crystals in cavities', from loose boulders in the gully just east of the granite – andesite junction at Le Côtil Pt. Anhedral crystals of 1cm cross the vein and extend several cm along it, and 'crystals with free faces' are up to 5mm in diameter.

Fig. 17

Fig. 18.

Fig. 19.

Fig. 20.

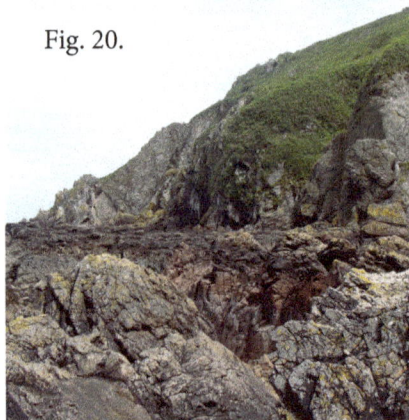

Mourant (1953, BSJ. pp. 18 - 19) in his article entitled 'The Garnets of Jersey' at the end of the Geological Report for 1952, describes the occurrence of 'chestnut brown garnets' associated with green patches of epidote (and other minerals) in the andesites, and cut 'cleanly across by veins of aplite', the main granite mass near the junction being almost an aplite.

Oliver (1958, BSJ. p. 181) records 'patches of massive garnet, chestnut brown in colour with epidote, quartz, chlorite & minor iron ore in metamorphosed andesite close to the granite contact at Côtil Point'. He also records that 'abundantly garnetiferous rock is found only as loose blocks at the foot of an inaccessible cliff'. The garnets are Andradite, Ca Fe rich, with Mourant (1953, p. 19) ascribing the Ca from 'amygdules' in the andesite, whereas Oliver thought it possibly derived from the granite along with the Fe.

The locations cited by those authors were examined several times and possible examples found, but weathered samples were uncertainly identified. Other possible examples were found later in the boulders of the pebble beach to the west. They occur in narrow elongate cavities in light grey, weathered and abraded, possible andesite, and though they match the description in colour (chestnut brown) and their shape or form seems to be naturally crystalline, identification is as yet uncertain as they could not be

Fig. 21.

Fig. 23.

Fig. 24.

Fig. 25. St. Saviour's Andesite, porphyritic.

extracted (Figs. 21, 22). However, the colour was thought to be superficial differential Fe staining of quartz as possibly with certain weathered samples (Fig. 23) (Dr. Hill pers. comm.)

The honey colour of foreign specimens (ref. the Internet) is shown below for comparison (Fig. 24).

Returning to the foot of the cliff path and the steep climb out, exploration of the beach pebbles in La Saline bay to the west reveals a variety of beach pebbles, both local, andesite and granite (Figs. 25, 26), and from further along the coast to the west, diorite and granite with xenoliths of diorite (Figs. 27, 28).

Brief Geological History.

After the deposition, uplift and folding of the Jersey Shale Formation, volcanic action produced the St. Saviour's Andesite disconformably on its eroded surface. This was followed by further volcanic activity when various types of volcanic rock, such as the Bonne Nuit Ignimbrite, flowed over the andesite and made up the St. John's Rhyolite Formation.

This period was followed by intrusion of the gabbros and diorites of Sorel and Ronez, and then followed by the NW granites of which the coarse St. Mary's granite and the Mont Mado type aplogranite of Bishop & Bisson (1989) were the most recent c. 450Ma ago.

Subsequent sea level changes produced the erosional features such as the 8m platform, stack and the cave while more recent powerful marine erosion as the sea level has risen, formed the narrow, steep - sided inlets eroded into the platform. Modern shingle deposits in the contact inlet have reduced the exposed height of the wave - cut platform.

Fig. 26. Mont Mado (micro) aplogranite.

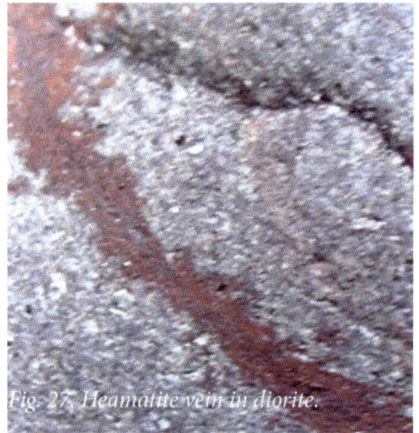

Fig. 27. Heamatite vein in diorite.

Fig. 28. Diorite xenoliths in granite.

References.

Bishop, A. C. & Bisson, G. 1989. Classical areas of British geology, Jersey. British Geological Survey HMSO.

Mourant, A. E. 1953. Garnets of Jersey. Ann. Bull. Soc. Jersiaise pp. 18 - 19.

Oliver, R. L. 1958. Andradite from the island of Jersey. Ann. Bull. Soc. Jersiaise. p. 181.

Robinson, A. J. & Mourant, A. E. 1936. Mem. de la Soc. Geologie & Mineralogie de Bretagne; t.3. p. 5-64). Not seen.

Robinson, A. J. 1938. Geology Section (in Rapport des Sections) for 1937. Ann. Bull. Soc. Jersiaise   p. 289.

Did you know...

That Ice Ages are caused by cyclic changes and wobbles in the tilt of the earth's axis relative to the sun, moving us closer to or further from the sun?

The Giffard Bay Trail

Volcanic rocks, lamprophyre dykes, raised beach.

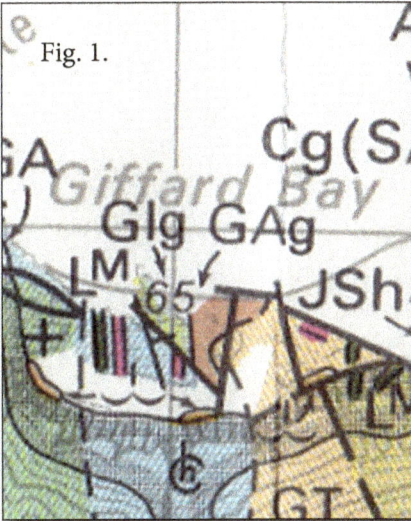

Fig. 1.

Giffard Bay is situated due east of Bonne Nuit Bay on the north coast of Jersey in St. John. It is best approached via the road from Bonne Nuit Bay eastwards and at about half way along, by branching off left via the narrow drive down and behind the Cheval Roc Apartments to La Crête Fort (Fig. 1). Park near the Fort and follow the cliff path east round the bay and c. 75m along, descend to the beach via a narrow path and steps.

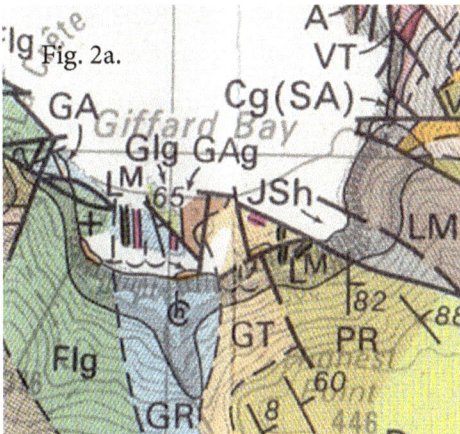

Fig. 2a.

KEY

Flg (Green). Frémont Ignimbrite (St. John's Rhyolite. Fm)

GA (Beige) Giffard Andesite (Bouley Rhyolite. Fm)

GR (Blue) Giffard Rhyolite (Bouley Rhyolite Fm)

LM (Green) Mica lamprophyre dykes (in Giffard Rhyolite).

(Red) Dolerite dykes.

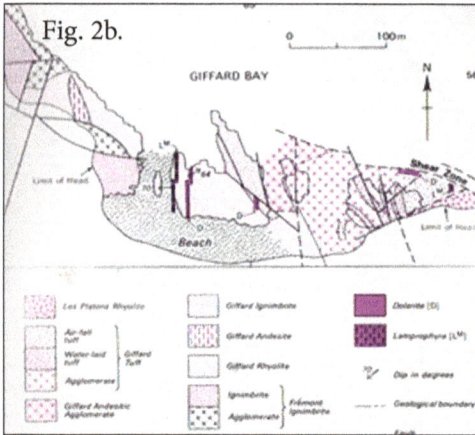

Fig. 2b.

GIFFARD BAY

0   100m

N 56

Sheer Zone

Limit of Head

Limit of Head

Beach

Les Platons Rhyolite | Giffard Ignimbrite | Dolerite (D)
Air-fall Tuff | Giffard Andesite | Lamprophyre (LM)
Water-laid tuff / Giffard Tuff | Giffard Rhyolite | Dip in degrees
Agglomerate | Ignimbrite } Frémont Ignimbrite
Giffard Andesitic Agglomerate | Agglomerate } | Geological boundary
| | Fault

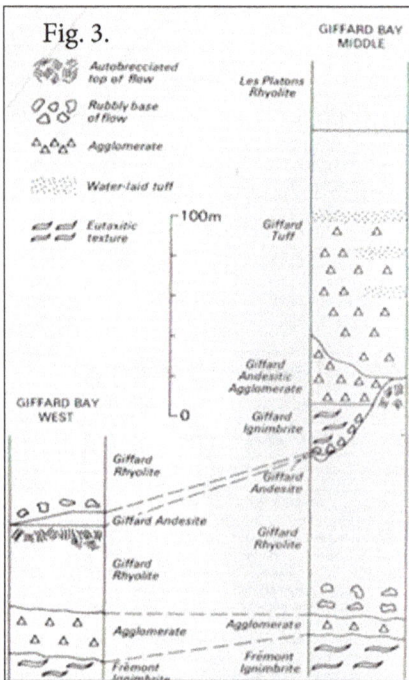

Fig. 3.

GIFFARD BAY MIDDLE

Autobrecciated top of flow
Rubbly base of flow
Agglomerate
Water-laid tuff
Eutaxitic texture

Les Platons Rhyolite

Giffard Tuff

100m

Giffard Andesitic Agglomerate

GIFFARD BAY WEST

0

Giffard Ignimbrite

Giffard Rhyolite

Giffard Andesite

Giffard Andesite

Giffard Rhyolite

Giffard Rhyolite

Agglomerate

Agglomerate

Frémont Ignimbrite

Frémont Ignimbrite

Fig. 4.

Fig. 6.

Fig. 5.

The trail starts on the beach rocks at the foot of this path which are the start of a variety of volcanic rocks ranging from ignimbrites to agglomerates and rhyolites with minor spherulites (Figs. 2a, b).

The Frémont Ignimbrite is the top unit of the St. John's Rhyolite Formation (base of left section, Fig. 3) and is exposed from La Crête Point below the path and

Fig. 7.

Fig. 8.

at the bottom of the steps onto the beach rocks where it is described as overturned based on the disposition of the flattened pumice and shards (eutaxitic texture) in the former tuff (Bishop & Bisson, 1989, p. 26) (Fig. 4).

Continuing across the rocks, the change to the Giffard Rhyolite occurs, it is brown to purple, finely crystalline and with areas of spherulites (Fig. 5) aligned along apparent flow - banding (Figs. 6, 7).

Patches of quartz crystals also occur, described as corroded and devitrified (op. cit. p. 31) (Fig. 8).

The beds exposed among the shingle, half way across the beach to the more central rock mass in the mid-low tide zone, show laminations similar to the flattened tuff layers (Figs. 9, 10). This seems to be in the position equivalent to the Giffard Andesite on the map but the structure seems more like that of an ignimbrite.

Towards the back of the beach there

Fig. 9.

Fig. 10.

Fig. 11.

Fig. 12.

Fig. 13.

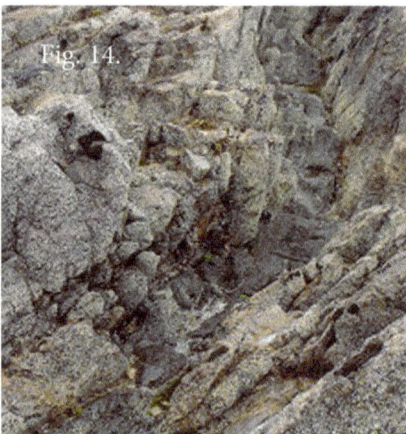

Fig. 14.

are outcrops of grey agglomerate, with angular clasts ranging from fine to coarse (>10cm) in unsorted exposures (Figs. 11, 12) to more uniformly size - sorted (Fig. 13). Many clasts are clast - supported while others are matrix - supported by a dark grey felsic groundmass. These outcrops may be equal to the Agglomerate between the Frémont Ignimbrite and the Giffard Rhyolite in the section (op. cit. p. 30), or to the Giffard Andesitic Agglomerate (not shown on the map), or to the western section, between the Giffard Rhyolite and Ignimbrite. However, the key on the map seems to indicate it is part of the Frémont Ignimbrite/Agglomerate, with the Giffard Andesitic Agglomerate cropping out further east.

The Trail then moves down beach to the central outcrop where three mica lamprophyre dykes can be seen. Two are shown on the map, but a third narrow one can be seen at the eastern

Fig. 15.

Fig. 16.

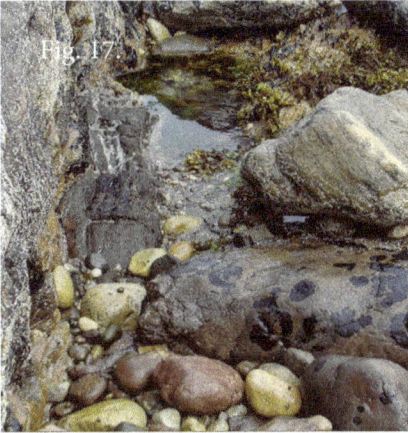
Fig. 17.

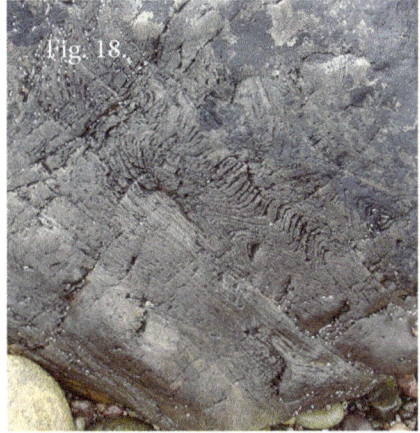
Fig. 18.

end. The dolerite one hasn't been found yet.

The dykes strike N - S and vary from c. 45cm wide (Figs. 14, 15) the western one having a sinuous outcrop, to c. 30cm wide with straighter outcrops for the eastern two (Figs. 16, 17). All are a faulted by dextral tear faults.

The mica lamprophyre dykes intrude the Giffard Rhyolite which is purple to brown and flow - banded, showing

Fig. 19.

Fig. 20.

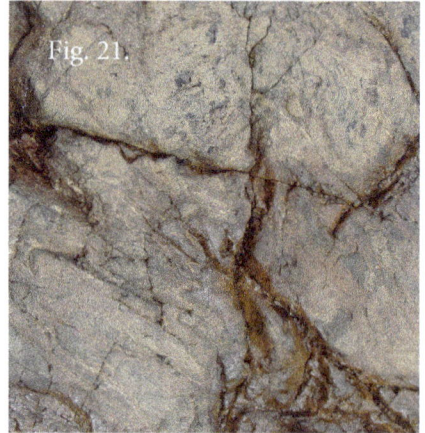
Fig. 21.

beautiful flow folds and laminar flow (Figs. 18, 19) individually.

Other units are brown to light brown and also show contorted flow - banding (Bishop & Bisson, 1989, p. 30). Laminar flow can still be seen picked out by more yellow bands and scattered spherulites also occur (Figs. 20, 21). No fully developed spherulitic layers were found here.

To the east on the cliff path, there are varieties of ignimbrite, recently described as rheomorphic or flow varieties (pers. comm. J. Sonnex, 2008, UCL unpubl. dissertation and G. Mason, UCL, unpubl. 2014). At present there is some discussion about the origin of the ignimbrite banding whether it is due to original flow or later compression (Figs. 22, 23).

In time sequence in this limited area, the Trail turns back to the cliffs at the

Fig. 22.

Fig. 23.

Fig. 24.

Fig. 25.

Fig. 26.

top of the beach. Here, there is a short section of the 8m raised beach cropping out at the foot of the cliffs westward. The pebbles are well - rounded and range in colour from light grey to brown, red, and purple with white bands set in a yellow - brown matrix (Figs. 24, 25). These are varieties of the local volcanic rocks, and rhyolite, jasper and flow - banded rhyolite can be seen (Fig. 26). Above, lies glacial head, a mixture of angular fragments in a partly loess - looking matrix.

Brief Geological History.

After the deposition, uplift and folding of the Jersey Shale Formation, the St. Saviour's Andesite volcanic rocks were deposited disconformably on the eroded surface. These were followed by rhyolite and ignimbrite volcanic rocks of the St. John's Rhyolite Formation seen at the start of this trail and the overlying rhyolites of the Bouley Rhyolite Formation. Here though, the rhyolites are interbedded with porphyritic and pyroclastic deposits. Further east in the bay, there are tuffs, and more agglomerates with pyroclasts of mudstone, andesite and rhyolite.

The environments of deposition in the bay area are discussed by Bishop & Bisson (op. cit. p. 31) who cite the work of Casimir & Henson (1955) and Thomas (1977) who identified a variety of terrestrial, fluviatile and lacustrine situations.

This is in contrast to the environments elsewhere, as recently the ignimbrites have been considered rheomorphic or flow types and there is much discussion about the origin of the banding in them, whether it is a result of flow or compression.

Subsequently, the volcanic rocks were uplifted and eroded to produce the Rozel Conglomerate to the east and the lamprophyre dykes were intruded. These occur with two different strikes, N - S here but NW - SE further west of Plémont where there is a minor swarm. This intrusive phase is later than the main dolerite phase of the Jersey Main Dyke Swarm (Lees, 1990) to the south.

There followed a period of faulting and then a very long period of weathering and erosion until the changes in sea level during the Ice Age produced a sequence of raised beaches during interglacials separated by Head and loess intervals deposited during glacial times. The 8m raised beach at Giffard Bay is part of the last transgression (interglacial) followed by the Head and loess of the last regression (glacial). The present Holocene sea level rise has caused some of the present day erosional and depositional landforms.

References.

Bishop, A. C. & Bisson, G. 1989. Classical areas of British geology; Jersey: description of 1:250,000 Channel Islands Sheet 2. London HMSO for British Geological Survey..

Brown, M. Power, G. M. Topley, C. G. & R. S. D'Lemos, R. S. 1990. Cadomian magmatism in the North Armorican Massif. p. 181 - 213. In The Cadomian Orogeny. Eds. D'Lemos, R. S., Strachan, R. A. & Topley, C. G., 1990, Geological Society Special Publication No. 51. Geological Society, London.

Casimir, M. & Henson, F. A. 1955. The volcanic and associated rocks of Giffard Bay, Jersey, Channel Islands. Proc. Geol. Assoc. Vol. 60. Pp 30 - 60.

Keen, D. H. 1978. The Pleistocene deposits of the Channel Islands. Rep. Inst.. Geol. Sci., No. 78/26.

Lees, G. J. 1990. The geochemical character of late Cadomian extensional magmatism in Jersey, Channel Islands, p. 273 - 291 in The Cadomian Orogeny, Geol. Soc. Spec. Publ. No. 51. (see D'Lemos et al. above).

Marett, R. R. 1911. Pleistocene Man in Jersey. Archaeologia, lxii, pp. 449-480.

Mourant, A. E. 1933. The Raised Beaches and Other Terraces of the Channel Islands. Geol. Mag. Vol. LXX, pp. 58-66.

Mourant, A.E. 1935. The Pleistocene deposits of Jersey. Bull.of the Soc. Jers.Vol.XII, pp. 489-496.

Ordnance Survey 1:250,000. Jersey. 2003. Official Leisure Map.

Renouf, J., James, L. 2010. High level shore features of Jersey (Channel Islands) and adjacent areas. Quaternary International (2010), doi: 10.1016/j.quaint.2010.07.005

Thomas, G. M., 1977. Volcanic rocks and their minor intrusive, eastern Jersey, Channel Islands. Unpubl. PhD thesis. Univ. of London.

Did you know...

That the sea - level falls during an Ice Age or Glacial Period, and rises during an Interglacial Period, one of which may have started when the ice melted c.10, 000 years ago?

The Gorey Harbour
– Petit Portelet Trail

Granite, Diorite, Shale, Lamprophyre dykes, Raised beach & Loess.

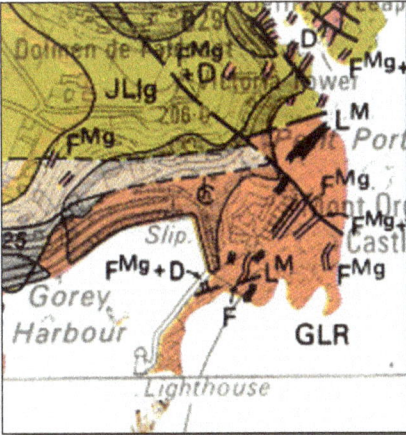

Fig. 1.

Gorey Harbour.

The Jersey Shale Formation in the west of Jersey with its various sedimentary structures and its contact with the NW granite at Le Pulec are well known, but the outcrops pictured below are parts of the Jersey Shale Formation which occur as far east as Gorey Harbour (Fig. 1) and show various other types of contact and structures. They crop out at the western end of Gorey Harbour north wall. The

Fig. 4.

Fig. 5.

Fig. 6.

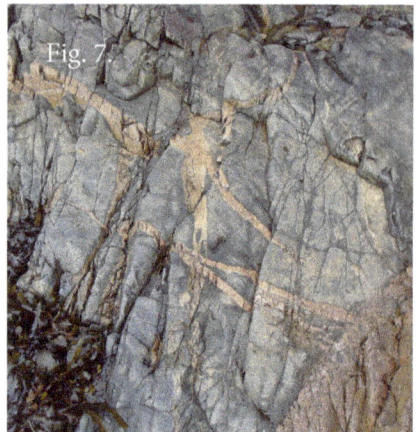
Fig. 7.

various structures and natures of their contacts are shown below.

Here, the shale is in contact with the Gorey granite, a browner, less quartzose variation of La Rocque granite of the SE granite (Bishop & Bisson, 1989, p. 57), and also with the black and white speckled diorites found on the south eastern coast in the SE Granite Complex (Brown et al, 1990, p. 196) from La Motte (Green Is.) to Seymour Tower.

The Jersey Shale Formation consists of dark grey, siltstone and greywacke, possibly metamorphosed to a hornfels near the contact with the granite. It has both laminations and interbeds of lighter coloured greywacke dipping NE, the outcrops occurring as enclaves of the Jersey Shale (Fig. 2). It is well - jointed and is intruded by granitic dykes and veins of different width and shape (Fig. 3), some birfurcating. There are two sets, one cutting the other nearly at right angles, demonstrating their differences in age (Figs. 4 & 5).

Yet others, have a ptygmatic (lobate) shape due to the varying competence of the host rock, and others are planar, and have formed by intrusion at different angles along various joint planes in the shale (hornfels) (Figs. 6, 7).

Some other narrow dykes have a uniform contact along one edge and an uneven contact with angular changes of direction along the other, suggesting very localised joint control during intrusion, while others show minor displacements due to small - scale faulting (Figs. 8, 9).

These granite and shale (greywacke) outcrops and relationships differ from those at Le Pulec because of the variety of their contacts, and because the granitic dykes and veins are also different, ie. there is an absence of felsitic, porphyritic and composite types (Figs. 10, 11). In addition, the

Fig. 8.

Fig. 9.

Fig. 10.

Fig. 11.

Fig. 12.

Fig. 13.

Fig. 14.

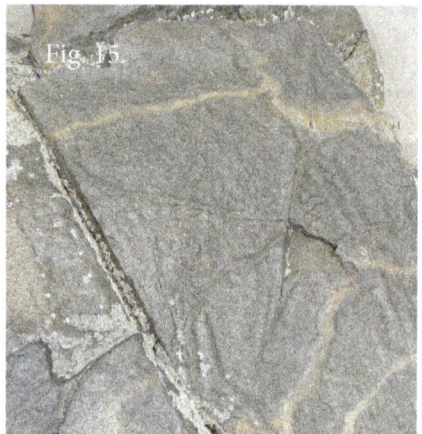

Fig. 15.

contacts here are more important because actual shale contacts with the SW granites are clearer than the one with the SW granite which occurs in Belcroute Bay and is described as crushed and grey near the contact (Bishop & Bisson, 1989, p.55), after the work of Henson, 1956, p. 266 - 295).

These outcrops allow one to see the nature of the contacts very clearly and the changes in texture in the rocks adjacent to them.

For example, here too, the shale has greywacke laminae and may also be a hornfels, having been thermally metamorphosed by the intrusion of granite veins and dykes of varying thickness(Figs. 12, 13).

The exposure of diorite occurs at the western end of the outcrop and is labelled

H on the IGS 1:25,000 map (1982).
It is more finely crystalline than the
diorite further south between La
Motte (Green Is.) and Seymour Tower
(Figs. 14, 15), but has a black and
white speckled appearance with the
white crystals (plagioclase feldspar)
being more yellow in parts. The black
hornblende crystals are also smaller
than those in the southern outcrops
and there are no large acicular
(needle - shaped) crystals or appinite
(porphyritic - like) textures.

Granite magma has intruded the
diorite here and forms narrow
undulating veins of varying widths,
and striking examples of small vugs
also occur one from a narrow feeder
vein and one with large white, quartz
and feldspar crystals and minor black
hornblendes (Figs. 16, 17).

The sequence of intrusion is shown
in one part of the outcrop where the
structures reveal the diorite intruding
the oldest Jersey Shale and the younger
granite veins intruding both the shale
and the diorite (Fig. 18).

Fig. 16.

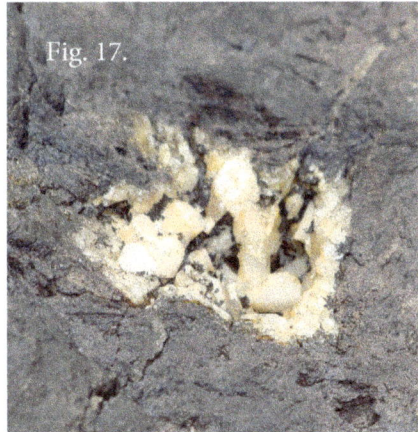

Fig. 17.

Petit Portelet bay.

A striking view of the Petit Portelet
beach area, with its two dykes, is
seen from the Castle battlements
and introduces this part of the trail
(Figs. 19, 20). The beach is reached
by returning to the steps up to the
harbour sea front, walking up the little
lane, La Petite Ruelle Muchie, crossing

Fig. 18

Fig. 19.

Fig. 20.

Fig. 21.

Fig. 22.

the Castle Green and descending the path to the beach.

The beach consists of a pebble strand at the back with sand forming a shore line area. Rock outcrops occur at the northern and southern ends with isolated masses in the middle. The rocks in the northern part of the bay form part of the St. John's Rhyolite Formation and consist of maroon ignimbrite, Jeffrey's Leap Ignimbrite. They are assumed to be faulted against the granite of the southern rocks which forms the Gorey section of the La Rocque granite, a part of the South-east granite (Bishop & Bisson, 1989, p. 57), or of the Southeast Granite Complex (Brown et al, 1990). This granite is browner in colour, less coarse and contains less quartz than around La Rocque.

Examination of the contact between the two rock types can be made in the centre of the bay in some of the

Fig. 23.

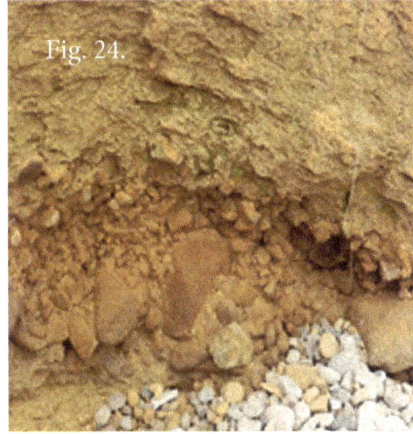
Fig. 24.

isolated exposures and seems to be without fracturing or brecciation.

Within the granite is a softer mica lamprophyre dyke which bifurcates and has been eroded into two gullies one of which strikes c. SW towards the sea wall (Fig. 21), whilst the other has been displaced by a small tear fault. Both branches contain a medium to dark brown mica lamprophyre with excellent medium to large bronze biotite mica crystals (Fig. 22).

There are striking yellow - brown cliffs at the back of the beach, which are predominantly made glacial head and loess. The erosion of a wave - cut notch at the base, together with the occasional land slip has revealed another example of a possible 8m raised beach, extending along much of the exposure (Figs. 23, 24), but also showing lateral variation and thinning (Fig. 25).

The pebbles vary in size, rounding and composition, an interesting type being andesite (Fig. 26), presumably from the exposure further west (IGS 1: 25,000 Channel Islands Sheet 2, 1978/82), as are the boulders, revealed at times within the pebble beach (Fig. 27). It is thought unlikely to be an outcrop as it does not appear below the rhyolite at the junction between the rhyolite and the granite further down beach, which is

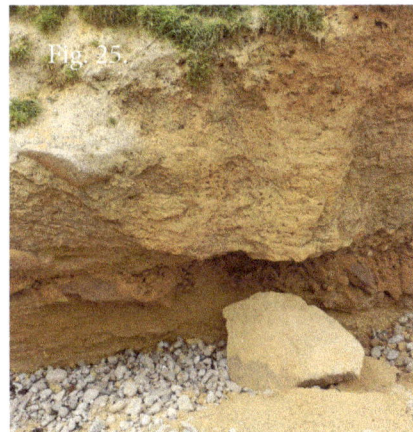
Fig. 25.

seemingly unfaulted and irregular in shape with possible xenoliths in the granite (Fig. 28).

Another interesting feature is a narrow vein of white to pink baryte (photo & pers. comm. Dr. A. Hill) striking SE - NW in a crevice and gully in the big headland at the northern end of the beach (Fig. 29).

Brief Geological History.

The Jersey Shale Formation is the oldest of the rock types and was deposited as a sequence of silts, sands and greywackes from turbidites in a deltaic environment. Named as Brioverian, it is upper Precambrian (Proterozoic) in age (c. 700Ma).

The shale formation was then uplifted and eroded. Vulcanism followed, producing the overlying andesites and rhyolites with further folding. The diorites were intruded during the following Cambrian period, and although the dates are unreliable, they seem to span a period during 570 – 550 Ma. These outcrops may be the northern part of the diorites and suspected gabbros cropping out along the south coast from St. Helier to Seymour Tower, separated from them by Le Hocq - La Rocque Granite and the Gorey Granite - Mont Orgueil Granite of the SE Granite Complex (Brown et al. 1990, p.196).

The red – pink granite then intruded

Fig. 26.

Fig. 27.

Fig. 28.

Fig. 29.

the shale and diorite units and also the rhyolites as 'there are a few inclusions of rhyolite' (Bishop & Bisson, 1989, p. 57), making it younger than them, but it is dated, again with some uncertainty, to 550 – 509 Ma (the possible age of the diorites). It was intruded in the way shown in these outcrops, possibly during a late folding phase of the Cadomian Orogeny. It is also considered to be a variation of the coarser La Rocque granite, with the usual felsic and mafic minerals, but with less quartz, and not as brown as the nearby granite of Mont Orgueil (Bishop & Bisson, 1989, p. 57).

There is no evidence of later Palaeozoic and Mesozoic rocks on the island although they occur nearby on the sea bed overlain by Eocene limestones.

The raised beach, loess and glacial head deposits, lying unconformably on the bed rock, represent a change from an interglacial period of high sea level, which produced the 8m raised beach when Jersey became an island, to a glacial period when the sea retreated, leaving it as a small plateau on a coastal plain, subject to cold, out - blowing, loess - laden winds from the northern ice sheet. The angular glacial head section represents a freeze-thaw or gelifluction deposit.

References.

Bishop, A. C. & Bisson, G. 1989. Classical areas of British geology. Jersey. Description of 1:25,000 Channel Islands Sheet 2. BGS. Her Majesty's Stationery Office, London.

Brown, G.M. 1978/82. Classical areas of British geology. Jersey. IGS Channel Islands Sheet 2. 1:25,000. Her Majesty's Stationery Office, London.

Brown, M., Power, G. M. et al. 1990. Cadomian magmatism in the North Armorican Massif. In The Cadomium Orogeny, p. 181 - 213. Geol. Soc. Spec. Pub. No. 51.

Henson, F. A. 1956. The geology of SW Jersey, Channel Islands. Proc. Geol. Assoc. Vol. 67. 266 - 295.

Nichols, R. A. H. and Hill, A. E. 2016. Jersey Geology Trail. Charonia Media. Lightning Source UK Ltd.

Did you know...

That this led to Jersey being an isolated plateau on a coastal plain when the sea retreated and to becoming an island when the sea advanced?

L' Étacq – Le Pulec Trail

Sandstone, shale & granite contact, & raised beaches.

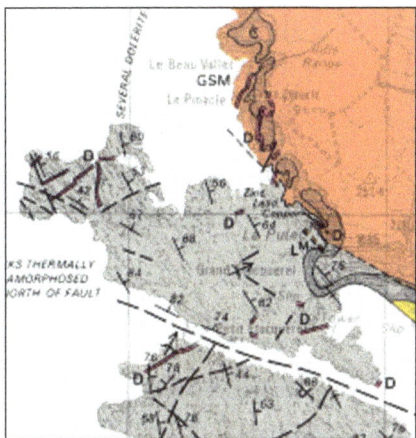

Fig. 1.

This is one of the best Trail routes because of the fine variety of features in well - exposed, accessible outcrops, and enables us to start with the first, or oldest rocks, at the base of the succession.

It is located at the northern end of St. Ouën's Bay (Fig. 1) and Le Grand Étacquerel and Le Petit Étacquerel are two prominent coastal land forms at L'Étacq (Fig. 2), while Le Pulec is an inlet just north of them.

The Trail is along outcrops of the Jersey Shale Formation and the NW granite with associated mineral veins in them, and access is easy.

Here, along with the best exposures of a main part of Jersey's geological history, the surrounding area, old cliffs, nearby quarries with mineralised areas, and a wave-cut platform to the west and south can also be explored to add more to the story.

Fig. 2.

Fig. 3.

Fig. 4.

Fig. 5.

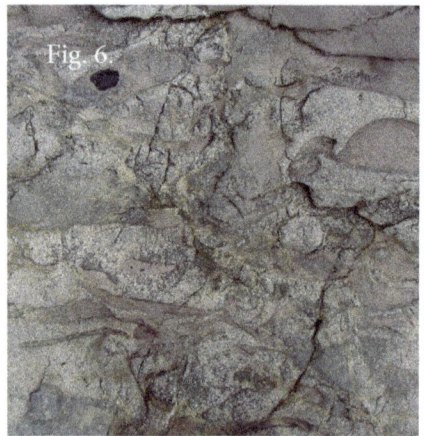

Fig. 6.

There is a wealth of sedimentary structures in the shale formation, which is more of a siltstone - greywacke sequence than a shale one, and there are several contact features with the granite, one of which includes a striking mineral vein. These are exposed in the landforms produced by marine erosion, which include former raised beach or gully features overlain by glacial head and loess.

It is better to start the Trail at L'Étacq, near Le Petit Étacquerel, and study the outcrops along the north side of the slipway down to the beach on the south side, to start with the facts before the interpretation.

Here, there is a sequence of thinly bedded, grey to brown, laminated siltstones, the laminae being light grey, medium-grained sandstones (turbidites or greywackes). The beauty here is in the variety of sedimentary structures which exhibit, cross-bedding, load-casting, attenuation or boudinage and balling (Figs. 3 - 6). Later micro - faulting and quartz veing can also be seen.

Returning to the top of the slipway, cross the road to the small quarry and look at the sole markings of the flute casts in the base of the overlying beds (Fig. 7).

Fig 7

From this quarry, cross the road and go down onto the pebbly beach and walk north to below the small carpark, west of the slipway down into Le Pulec (or Stinky Cove if there's a lot of rotting seaweed in it). About 50m along the beach outcrops of Jersey Shale reveal beautiful ripple - laminated structures on the bedding plane surface (Fig. 8). At first sight they appear like shallow water beach ripples but cross - sections show the cross laminae of turbidity current deposits.

Fig 8

Above this in the low outcrop, the beds reveal clear sections of cross - laminated intervals c. 1 - 2cm thick (Figs. 9, 10). The cross laminae of coarser sandstone or greywacke, c. 1mm thick are very clearly exposed and show current bedding as well as grading.

Other beds with lobate bases, show evidence of scouring as cut-and-fill structures, and of compression to give load casts and to have caused thickening and thinning. Incipient sedimentary dykes can also be seen, as well as later jointing and displacement by small faults. Immediately to the north west, the low cliffs reveal beds dipping landwards to the northeast.

Retrace one's steps or scramble up the low cliff to the small car park. From here, the view across the inlet, Le Pulec, reveals higher beds of the grey Jersey

Fig. 9.

Fig. 10.

Fig. 11.

Shale Formation in section, the beds dipping away from you eastwards, and with light - coloured veins and thin dykes crossing them (Fig. 11). Above are cliffs of the NW granite rising steeply to Les Landes, the flat - topped plateau of northwest Jersey.

The cliffs also extend NW and the junction with the Jersey Shale Formation can be seen extending irregularly along the beach rocks of a narrow wave - cut base at low water. Small masses of light brown - yellow granite, like wide dykes, are separated by outcrops of the dark grey shale (Fig. 12). Faults and master joints can be easily seen in the cliffs above, marked by caves in several cases.

The Trail continues into the inlet down the slipway cut through the shale beds which reveals their dip and some faults with crushed rock zones on the left side, and other more recent features which can be studied on the way back in keeping with the stratigraphic order of their formation.

Fig. 12.

Fig. 13.

Fig. 14.

Fig. 15.

Fig. 16.

If the inlet is relatively clear of seaweed the cliffs and low beach rocks can be easily examined for bedding features and intrusive veins, thin dykes and minerals.

Near the base of the Jersey Shale sequence there are good exposures of various veins and thin felsite dykes of differing structure and composition, ranging from planar to transgressive and from single to composite with thin and occasional flame - like offshoots (Figs. 13 - 16).

The junctions are very well defined even of the offshoots.

They vary from grey to grey - pink in colour and from 2 - 4cm wide or thick, and from medium crystalline quartz and feldspar to porphyritic with larger

Fig. 17

Fig. 18

feldspar crystals. The outer, and in some parts, the inner edges of the composite central part are granitic in composition and texture. Two phases of intrusion are visible along with minor faulting.

The inlet floor is covered with a mixed pebble deposit of shale and granite pebbles with low rock exposures cropping out. It is in this area on the western side that a vein of silver, lead and zinc was found and surveyed in the 1800s but found to be uneconomic to mine. At present, some may be found but much is covered by seaweed and shingle.

However, in the beach rocks cropping out in the floor on the eastern side of the inlet among the pebbles, are examples of quartz veins with small, scattered pyrite crystals. The iron pyrite crystals are c. 1mm in size and some are octohedra (Figs. 17, 18). Careful examination may reveal more in other veins.

Continuing further NW to where the cliffs finish and a low rock platform extends to the main cliff base, scramble onto the platform, and cross to the junction between the shale and the granite which exhibits excellent vein intrusions from the granite (Fig. 19, 20) and where the shale has been contact - metamorphosed to a hornfels with concentrations of chlorite and minor cordierite and biotite leading to spotted hornfels, north and south of Le Petit Étacquerel. The granite is a coarsely crystalline pink to orange, quartz and orthoclase feldspar garnite with minor biotite and hornblende and many aplite veins.

This part of the trail will lead you to a narrow gully eroded along a zinc blende vein with the sphalerite set in a gangue of yellow ankerite (Fe Mg limestone). Along the gully the various relationships of the ore mineral to the gangue

Fig. 19.

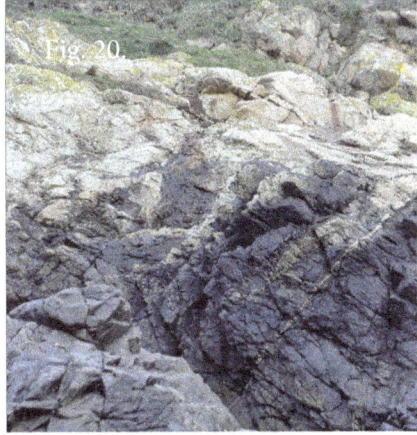
Fig. 20.

(uneconomic) mineral are strikingly displayed.

Textures, showing the disposition of the sphalerite and the ankerite, vary from filigree - like to brecciated, and narrow linear sphalerite within and along the edges of the ankerite, and to large and small concentrations (Figs. 21 - 24). The nearest exposure to the silver, lead vein on the western side of the gully, shows a quartz gangue within the shale, with dark grey

Fig. 21.

Fig. 22.

Fig. 23.

Fig. 24.

Fig. 25.

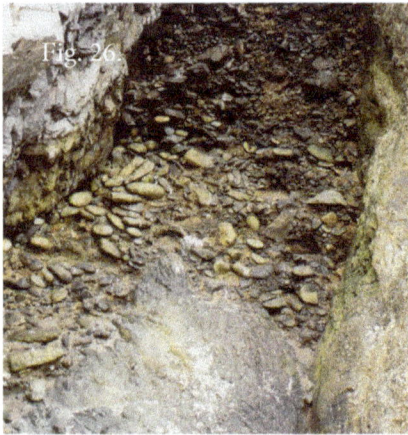
Fig. 26.

minerals that have not been tested for their type (Fig. 25).

From here, the Trail leads back to the slipway to examine the south side of the cutting where two gullies filled with rounded pebbles are overlain by Pleistocene glacial head and loess, and indicate remnants of raised beaches or raised - beach gullies; a third one occurs in the rocky cliffs on the west side of the inlet (Figs. 26 - 28).

Fig. 27.

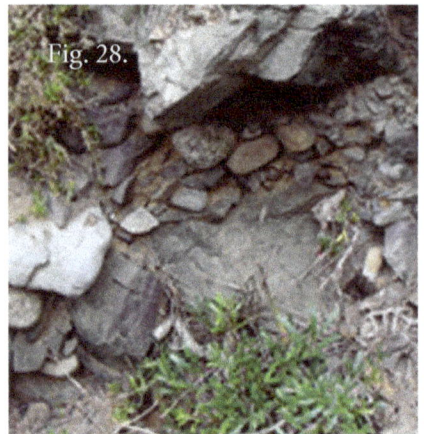
Fig. 28.

The pebbles show poor to average size-sorting and are well rounded to sub - spherical and disc-shaped clasts, varying between clast - supported and matrix - supported by a yellow - brown silt to sand groundmass, possibly loessic in part.

The channels, seemingly gullies eroded in the 8m wave - cut platform (similar to the ones on the foreshore today) vary in width and depth, one being deep and narrow and the others being wider and shallower. The pebble beds are overlain by a thick sequence of angular fragments of shale showing incipient layering. This is considered glacial head formed during freeze - thaw conditions with soil creep or slumping due to gelifluction.

Brief Geological History.

The rocks of the Jersey Shale Formation have been mapped in detail and more recently have been divided into four facies within a deltaic (submarine fan) environment by analogy with the work of several researchers (Bishop & Bisson, 1989, p. 8 - 10, who quote the work of others, see References).

Around the trail area and on the eastern edge, has the most important associations of ripple laminations, cross - stratification and grading; these are parts of mid to outer fan deposits. Palaeocurrent directions are generally northwards and the shales young towards the east where they are overlain disconformably by a sequence of andesites and rhyolites.

No fossils have yet been found but they are dated as Precambrian, c. 700 Ma old and have been folded at four different times with fold axes and plunging folds being exposed in several places south of L'Étacq, and later faulting produced varied displacements, the largest being along La Bouque Fault just to the south.

Volcanic activity occurred in the east followed by uplift and erosion to produce the Rozel Conglomerate Formation.

The area was intruded by the Northwest Granite Complex dated at c. 475 Ma and this gave rise to the mineralisation. The area then underwent a very long period of erosion which produced the extensive wave - cut platform to the west and the fossil stacks of L'Étacq and St. Ouën's Bay to the south.

The last changes in sea level brought about by the Ice Age glaciations finally came to an end after producing the 8m wave - cut platforms, stacks and the raised beaches during advances, and the loess - head deposits during the retreats seen in the last part of the Trail.

References.

Bishop, A. C. & Bisson, G. 1989. Classical areas of British geology; Jersey: description of 1:250,000 Channel Islands Sheet 2. London HMSO for British Geological Survey..

Brown, M. Power, G. M. Topley, C. G. & R. S. D'Lemos, R. S. 1990. Cadomian magmatism in the North Armorican Massif. p. 181 - 213. In The Cadomian Orogeny. Eds. D'Lemos, R. S., Strachan, R. A. & Topley, C. G., 1990, Geological Society Special Publication No. 51. Geological Society, London.

Keen, D. H. 1978. The Pleistocene deposits of the Channel Islands. Rep. Inst.. Geol. Sci., No. 78/26.

Lees, G. J. 1990. The geochemical character of late Cadomian extensional magmatism in Jersey, Channel Islands, p. 273 - 291. In The Cadomian Orogeny, Geol. Soc. Spec. Publ. No. 51. (see D'Lemos et al. above).

Marett, R. R. 1911. Pleistocene Man in Jersey. Archaeologia, lxii, pp. 449-480.

Mourant, A. E. 1933. The Raised Beaches and Other Terraces of the Channel Islands. Geol. Mag. Vol. LXX, pp. 58-66.

Mourant, A.E. 1935. The Pleistocene deposits of Jersey. Bull.of the Soc. Jers.Vol.XII, pp. 489-496.

Renouf, J., James, L. 2010. High level shore features of Jersey (Channel Islands) and adjacent areas. Quaternary International (2010), doi: 10.1016/j.quaint.2010.07.005

Nichols, R. A. H. and Hill, A. E. 2016. Jersey Geology Trail. Charonia Media. Lightning Source UK Ltd.

Did you know...

That the sea level has changed three times and that Jersey has raised beaches at 8m, 18m and 30 - 40m above present sea level, and fossil coast - line cliffs inland?

The Plémont Trail.

Coastal inlets, caves, arches, stacks, reefs...& a Minor Dyke Swarm.

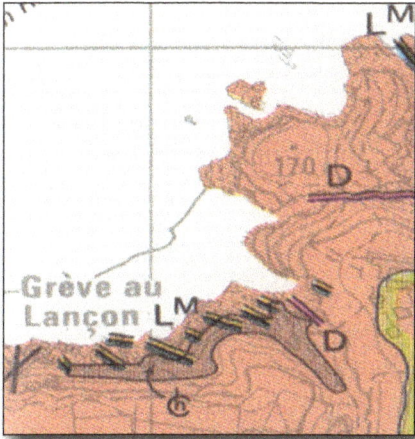

Fig. 1.

La Grève au Lanchon (Lançon en Jèrriais) is a bay to the west of La Tête de Plémont situated on the northwest coast of Jersey in St. Ouën (Fig. 1).

It is backed by granite cliffs in which there is a noticeable concentration of mica lamprophyre dykes, 9 in 400m (green, Fig. 2) with one dolerite, whereas others crop out singly, eg. only two westwards to Gronez, and in very minor numbers c.1km apart eastwards along the rest of the coast.

Additionally, there are veins and xenoliths of several types of rock such as aplite, dolerite (red, Fig. 2) and diorite, faults and many joints and master joints (black lines, Fig. 2).

There is also a striking variety of land forms such as caves, inlets, arches, stacks and reefs, caused by differential marine erosion along zones of softer rock in the pink granite.

The pink - red, well - jointed granite is named the North - west granite, St. Mary's type (Bishop & Bisson, 1989,

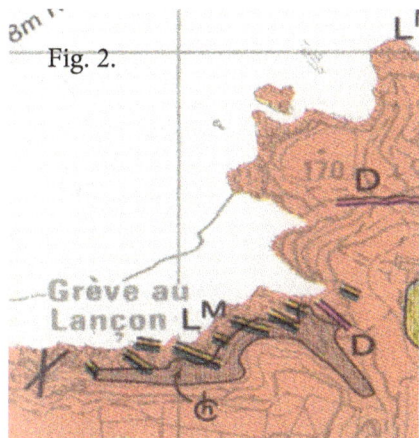

Fig. 2.

p. 52), or the Porphyritic Granite, one of four different types making up the Northwest Granite Complex (Brown et al. 1990, p. 195).

It is called porphyritic because it contains large zoned and unzoned orthoclase feldspars, in a finer crystalline groundmass together with scattered black biotite and hornblende crystals. Inclusions (xenoliths) of dark grey and white speckled diorite of variable size and shape also occur.

Glossary.
The following words are used to describe the rock types, land forms and structures.

Aplite	*A pink - orange, finely crystalline igneous dyke or vein rock of quartz and feldspar.*
Dolerite	*A grey, medium crystalline igneous dyke rock rich in iron and magnesium.*
Dyke	*A wall - like, discordant structure, usually softer, dark grey dolerite, within pink granite.*
Fault	*A fracture caused by tension or compression which has displaced adjacent rocks.*
Granite	*A coarsely crystalline pink igneous rock rich in quartz, feldspar and micas.*
Joint	*A fracture caused by tension in the rocks but without any displacement.*
Lamprophyre	*A brown, medium crystalline, basic to ultra basic (ultra mafic) igneous, minor intrusive dyke rock with large brown mica crystals; the magma is derived from depth (Gk. lampros = bright; phyro = to mix).*
Vein	*A narrow, discordant structure of crystals cutting through the country rock.*
Xenolith	*A different rock type surrounded by the country rock.*

The East side Story.

The Trail starts east of the bottom of the steps down to the beach called La Grève au Lanchon (sand eel beach) and in the furthest east of two inlets.

The east side is dominated by landforms caused by marine erosion and modified by mechanical and chemical weathering. Well - jointed cliffs (Fig. 3) of granite form the eastern wall of the first inlet. Here, nearer the northern part, the pink to brown porphyritic granite contains a large, lozenge - shaped xenolith of dark grey diorite, the edges of which vary from well-defined

Fig. 3.

Fig. 4.

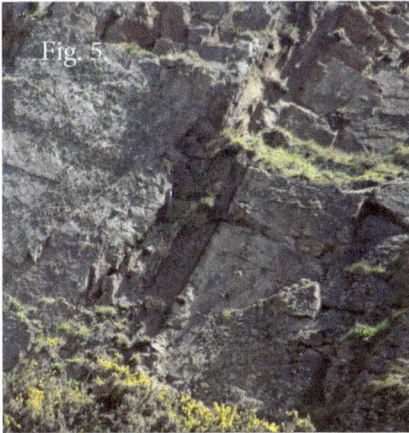
Fig. 5.

to ragged and show inclusions of pink feldspar crystals, possibly during assimilation of the older diorite (Fig. 4). Similar xenoliths of smaller size and variable shape occur in the granite beach rocks west of the steps (see below).

This first inlet is a wider and longer cove which extends further south due to erosion along a dyke, seemingly a mica lamprophyre (from its colour), and along many joints seen in the back wall (Fig. 5).

Returning towards the steps into the second, smaller, narrower inlet, one can see low reefs in the beach, several stacks and two narrow arches, all leading south to a large cave in the back wall. Here, the two arches lead westwards into the third inlet (the first one east of the steps). These features are classic examples of erosion along cross - joints from one inlet to another to produce a cave then an arch.

The arch (Fig. 6) occurs over a long, vertically - sided narrow cleft open at both ends, which has been eroded along a fault zone of softer, crushed rock, leaving an arch of crushed rock Note the darker red, iron - rich vein in the arch and that the shapes of the landforms are controlled by the fracture planes.

Fig. 6.

The stack, or pinnacle rock (Fig. 7), is caused by erosion along vertical joints approximately at right angles to each other, leaving a pinnacle with flat sides which are the joint planes.

The cave at the back has been eroded along a well - weathered mica lamprophyre dyke seen in the granite some metres in front of the cave and in the cliff face above the cave (Figs. 8 & 9) which strikes at 170° and cuts across inclined joints, so is therefore later; it also bifurcates in the roof.

These inlets, like the first one, have been eroded along dykes, joints and faults

Further weathering and stream erosion from above will produce gullies in the land surface. The arch will collapse leaving a stack and a new cliff face. The stacks will then be eroded by wave action and collapse, leaving low reefs. In this way, land is lost by cliff recession.

The West Side Story!

The features on the west side are dominated by cliffs and caves which have

Fig. 8.

Fig. 9.

Fig. 10.

Fig. 11.

Fig. 12.

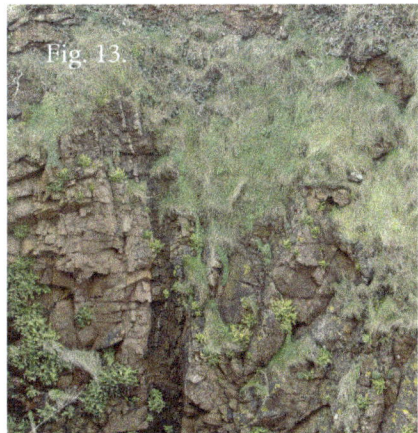
Fig. 13.

been eroded along fractures and softer rocks, and by low cliffs and reefs separated by gullies (Fig. 10).

As you walk down the steps to the beach, there is a deep, vertical - sided inlet on your left which is being eroded by stream action at the back as well as by wave action from the beach. In this inlet immediately west of steps, remnants of a dolerite dyke can still be seen stuck to the eastern wall of the gully at beach level (Fig. 11) although none can be seen further into the gully or in the back wall. The dolerite dyke strikes at 125° and only 80 - 100cm are visible in an 8m wide inlet, and it has a similar strike to that of the mica lamprophyres.

However, in the back wall under the waterfall, the rock is heavily stained and a mica lamprophyre dyke can be seen which ascends the western wall (Fig. 12) and its upper part can be seen on the opposite wall from the steps (Fig. 13).

A cave has been eroded below it, but opposite it, is the second arch which goes through to the inlet east of the steps (Fig. 14.)

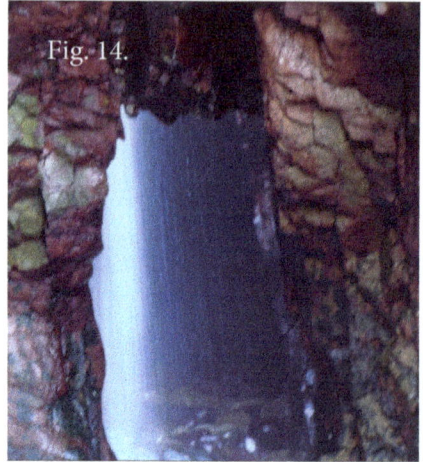

Fig. 14.

Meanwhile, in the granite outcrops on the beach, there are areas of dark grey and white speckled diorite some with isolated feldspars indicating partial alteration by the granite. These are inclusions of older rock incorporated by the later granitic magma during its intrusion (Figs. 15, 16). They may represent relict parts (xenoliths) of the diorites and gabbros exposed further east at Sorel and Ronez Points. Additionally, in the first cliff outcrop to the west, there are three main directions of the joint planes, not always at right angles to each other; the planes dipping towards the beach could cause potentially dangerous rockfalls.

Fifty metres west, a gully striking SW – NE leads to the first cave and has been produced by erosion along a near vertical master joint plane.

On the east side of the gully are two narrow brown dykes going up through the granite, striking at 125°; these are rarer than dolerite dykes and are mica lamprophyres. The first one, showing displacement, has been faulted, and the second one with chilled margins has been intruded into the colder granite. Alternatively, it is a multiple intrusion, that is, one dyke intruded by another (Fig. 17, 18).

Further SW towards the cave, the second dyke, a well - jointed, multiple lamprophyre with lenses of granite occurs striking with the same strike (Fig. 19). To the west, this dyke continues through the granite (Fig. 20).

The first cave in the cliffs has a narrow, high entrance, 2 - 3m wide, and the front face and initial part of the roof can be easily seen (Figs. 21, 22). It strikes SSW into the cliff (in contrast to the ESE strike of the dykes) and appears to have been eroded along a fault and or master joints as there is no evidence of a dyke.

In the next cove, the second cave west is a trapezoid - shaped one, and here again, the strike of the cave is different from that of the dykes and it appears to

Fig. 15.

Fig. 16.

Fig. 17.

Fig. 18.

Fig. 19.

Fig. 20.

Fig. 21.

Fig. 22.

have been eroded along and between a master joint and a fault on the right hand side (Figs. 23, 24).

Raised beach deposits may occur above these caves. Four mica lamprophyre dykes crop out in the eastern wall of the gully leading to the second cave. They all have a similar strike to c. 125° and vary in width from 15cm to 1m. They are sinuous in parts and in some cases have bifurcated. They truncate the aplites which show limited displacement as if there has been minimal faulting during intrusion (Figs. 25 – 29).

In this cove, the first and second dykes show the bifurcation and sinuosity (Figs. 25, 26) while the third mica lamprophyre dyke on the eastern side curves SE round to S in its upper part, and possibly shows multiple intrusive phases (Fig. 27). In its lower part, the dyke bifurcates and attenuates following a joint at the base (Fig. 28).

Fig. 23.

Fig. 24.

The fourth (Fig. 29) is similar to the first two and clearly bisects two orange aplites.

However, in the third cove towards the main cliff line with overhangs rather than caves, a wide gully strikes back E and from this a side gully strikes south to the cliff line. On the eastern wall of this gully is another lamprophyre dyke striking in a similar direction to the others (Fig. 30).

Fig. 25.

Fig. 26.

Fig. 27.

Fig. 28.

Fig. 29.

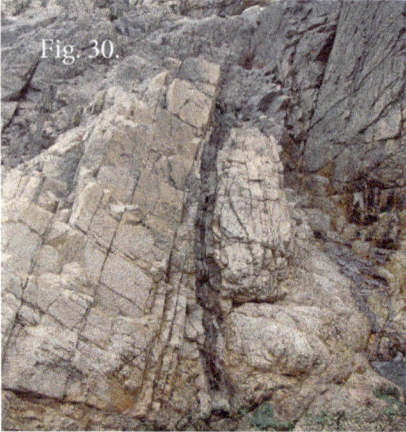

Fig. 30.

The orange dykes are aplites (also called microgranites), acid intrusions of finely crystalline equigranular quartz and feldspar, noticeable because of their colour, strike and dip, and their age. They are rich in iron oxide and have a light selvage at the junction with the granite. They strike both sub - parallel to (Fig. 31, 32), and across the basic dykes which bisect them, and are either vertical or inclined, dipping to the east. They intrude the granite but are themselves cut and sometimes

Fig. 31.

Fig. 32.

Fig. 33.

Fig. 34.

Fig. 35.

Fig. 36.

displaced by the dolerites and lamprophyres (Fig. 33), and are therefore older than the latter.

Finally, climbing the steps from the beach, binoculars are needed to examine an exposure of Head above a dyke at the top of the inlet cliff to the west in which among angular fragments there may be some rounded ones which may represent part of a raised beach (Figs. 34, 35), while to the NE, perched in the cliffs there may be an 18m a sea cave (Fig. 36).

Brief Geological History.

During Lower Palaeozoic times, the various granites of the Northwest Granite Complex formed as follows, the Porphyritic granite (c. 465 Ma) formed first and was followed by the Coarse Granite (c. 438 Ma) and the Mont Mado microgranite or Red Granite (c. 426 Ma). They intruded the Precambrian Jersey Shale Formation after its deposition, uplift and folding, and during the eruption of the andesites forming the lower part of the Volcanic series (c. 522 – 477 Ma).

These granites are younger than the adjacent Southwest Granite Complex (c. 550 - 483 Ma) with which they are shown to be connected (Bishop & Bisson, 1989, p. 101 - 2), and younger than the 483 Ma age for the dolerite dykes of the Main Dyke Swarm intruded along E - W strikes on the south coast. The aplite dykes were then intruded cutting cleanly through the granite.

The N - S dolerite dykes were intruded, possibly at the same time as the N - S tear-faulting and they cut the earlier E - W dykes. The mica lamprophyre dykes with their variable strike of NE at Gorey to NW at Plémont are younger again and were intruded last, as seen by their cross - cutting relationships with the dolerite dykes.

Intrusion by the NW granites also occurred during the eruption of the rhyolites and the uplift, erosion and deposition of the Rozel Conglomerate (c. 477 – 426Ma).

The folding and intrusion during the Precambrian and Lower Palaeozoic occurred during a period from c. 700 – 425 Ma, ie. during 275 Ma from the Precambrian to Silurian, a long period known as the Cadomian Orogeny (Brown et al, 1990, p. 181 et seq.).

This seems to have been followed by a long period of erosion, during the Upper Palaeozoic and Mesozoic, which removed the country rock and revealed the granites, until Tertiary limestones were deposited around the island during Eocene times.

During the Pleistocene there were several periods of higher sea level during interglacial times when raised beach were formed, the 30m one being the oldest; other raised beaches deposits occur at 8m and 18m. These were interspersed with periods of loess and head deposition on the eroded granite bedrock, as seen at the top of the cliffs in this area, during intervening glacial times. The present climatic regimes and weather, which control the present weathering, marine and fluvial erosion and deposition, have produced the land forms and the beach sands illustrated above.

References.

Bishop, A. C. & Bisson, G. 1989. *Classical areas of British geology; Jersey: description of 1:250,000 Channel Islands Sheet 2. London HMSO for British Geological Survey.*

Brown, M. Power, G. M. Topley, C. G. & R. S. D'Lemos, R. S. 1990. *Cadomian magmatism in the North Armorican Massif. p. 181 - 213. In The Cadomian Orogeny. Eds. D'Lemos, R. S., Strachan, R. A. & Topley, C. G., 1990, Geological Society Special Publication No. 51. Geological Society, London.*

Keen, D. H. 1978. *The Pleistocene deposits of the Channel Islands. Rep. Inst.. Geol.Sci., No. 78/26*

Lees, G. J. 1990. *The geochemical character of late Cadomian extensional magmatism in Jersey, Channel Islands, p. 273 - 291. In The Cadomian Orogeny, Geol. Soc. Spec. Publ. No. 51. (see D'Lemos et al. above).*

Marett, R. R. 1911. *Pleistocene Man in Jersey. Archaeologia, lxii, pp. 449-480.*

Mourant, A. E. 1933. *The Raised Beaches and Other Terraces of the Channel Islands. Geol. Mag. Vol. LXX, pp. 58-66.*

Renouf, J. 1986. *Geological setting and origin of La Cotte de St. Brelade, p. 35-52, in (Eds.) P. Callow & J. Cornford. La Cotte de St. Brelade: 1961-1978, Excavations by C.B.M. McBurney.*

Renouf, J., James, L. *High level shore features of Jersey (Channel Islands) and adjacent areas. Quaternary International (2010), doi: 10.1016/j.quaint.2010.07.005*

Stevens, C., Arthur, J., & Stevens, J. 1985. *Jersey Place Names, Vols. I & II. Société Jersiaise.*

Did you know...

That Jersey's oldest siltstones and sandstones, seen at L'Étacq, formed from turbidity currents on the slopes and in the channels of a submarine delta more than 600 Ma ago?

Fig. 2.

The Portelet Bay Trail

Raised beach, loess & head, dykes & sills, & beach pebbles.

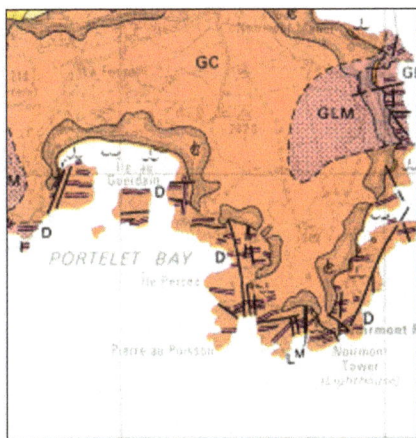

Fig. 1.

Portelet Bay (Fig. 1) was revisited because there are other prospects with interesting geological features, sometimes overshadowed by the cliffs with their striking and well - known 8m raised beach, loess and head. During a previous visit, certain features were either not exposed, visited landforms so Section members recorded their observations of the other features.

Looking south from the top of the steps to the beach in Portelet Bay, there is a striking view of the red, coarsely crystalline Corbière granite, the outermost of the three granites forming the Southwest Granite Complex (Brown et al. 1990). The granite cliffs enclosing the bay extend to the east and west and shelter the central Île au Guerdain (Janvrin's Tomb). The cliffs are steep, fractured by joints, the planes of which often form the sheer cliff faces, and are intruded by dolerite dykes, the erosion of which has left deep gullies cutting into the cliffs (Fig. 2).

At first sight, the other geological prospects and features aren't evident, but a small wave - cut platform and notch at base of the western headland and several deep, narrow gullies or clefts to the east seem to indicate a varied geological history, illustrated by those different views or prospects.

These other prospects are indications of several other geological events of which the raised beach, its superficial deposits and the present bay are the

Fig. 3.

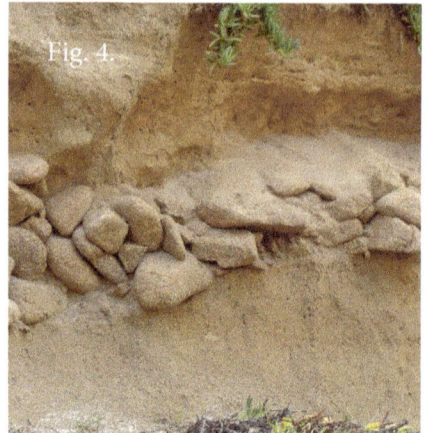
Fig. 4.

most recent.

Down on the beach, the cliffs at the back are strikingly yellow - brown and composed of a former beach strand c. 50cm thick, of large, rounded granite pebbles of variable size and various beds of sand, loess and head.

The sequence lies unconformably on an uneven, eroded granite bedrock surface at present day beach level (Fig. 3) and the lowest part of the section is a red - brown clay, possibly loess. The raised beach is composed of large, rounded to sub - rounded, pink, coarsely crystalline granite boulders with a low degree of sphericity (Fig. 4). The thickness of the deposit varies around 50cm but thicker 'pockets' also occur (Fig. 5). It dips gently from east to west along the cliff line and rises to the eastern side of the steps.

Fig. 5.

Fig. 6.

Fig. 7.

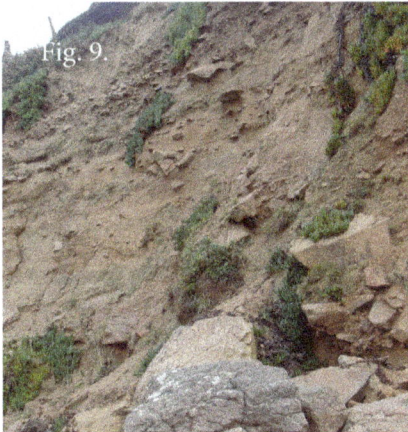

Fig. 8.

Fig. 9.

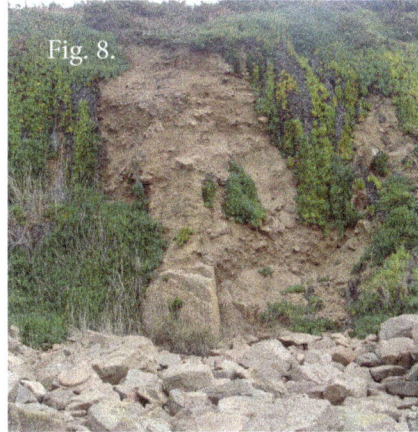

Fig. 10.

It is overlain by a light, red - brown sand, aeolian or marine in origin, weathered and eroded into hollows (Fig. 6). This is in turn is overlain by glacial head with angular fragments, and finally by a yellow, clay - silt (loess, a wind - blown deposit) (Fig. 7), parts of which contain angular granite blocks forming glacial head (a freeze - thaw deposit due to gelifluction).

However, following the projected position east of the steps, no present evidence of raised beach could be found, but thick deposits of loess and head are exposed theoretically equivalent to those deposits further west (Fig. 8 - 10) and at the eastern end of the cliff section of thick loess and head, there are exposures of variable thickness, of a bed of rounded pebbles between large angular boulders, resting with a horizontal base on loess which overlies an eroded granite bedrock surface.

As a tribute to Arthur Mourant, a famous Jersey geologist, his observations made in 1932 - 3 are included below (Mourant, 1933, p. 61);

5. Yellow Clay (loess) . . . . 10' 0" (c. 3.00m)
4. Reddish brown stony clay (head) . . . . 10' 0" (c. 3.00m)
3. Reddish brown sand . . . . 5' 0" (c. 1.60m)
2. Large well rounded pebbles . . . . 1' 6" (c. 0.45m)
1. Reddish brown clay . . . . 2' 0" (c. 0.60m)
Lowest deposit visible.

Respectively, as described by several other authors, the deposits represent a glacial period deposition for the lowest deposits, with the 8m raised beach, a higher sea - level deposit, representing warmer interglacial conditions. The overlying loess and head deposits represent drier glacial period deposits again, deposited by out - blowing winds and under freeze - thaw conditions respectively.

Further west along the beach, other prospects are revealed. There are many boulders of different sizes and of different rock types. The granite boulders reveal several different textures also found within the granite cliffs further west. There are those with fine to coarse crystalline textures and porphyritic textures (Figs. 11, 12), and those showing variable concentrations of minerals which produce pegmatite textures (Fig 13). All of these show the varied modes of crystallisation during the cooling of the granite.

In addition, grey shale pebbles show some of their sedimentary and tectonic history in the shape of possible slumped laminae and quartz vein minor folds (Fig. 14).

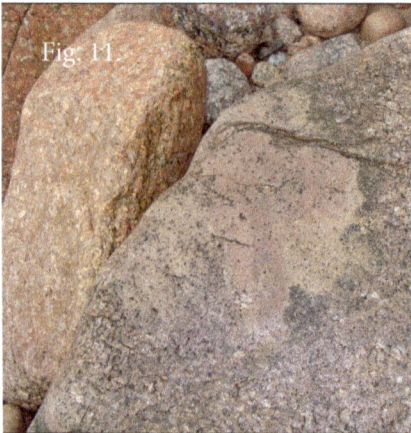

Fig. 11.

Fig. 12.

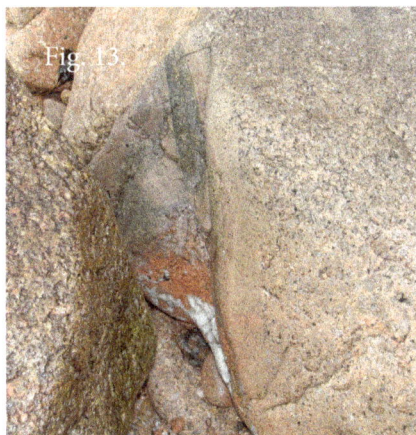

Unexpectedly, there are also well - rounded pebbles of black and white speckled diorite, some with pink feldspars, grey and white porphyritic andesite, grey quartz - feldspar porphyries and maroon spherulitic rhyolites, some with very small white crenulated spherulites.

The nearest outcrops of similar diorite and andesite occur around Elizabeth Castle and immediately west of St. Helier respectively while outcrops of the porphyritic andesite occur in the quarries just to the north.

However, the nearest outcrops of the maroon types of spherultic rhyolites crop out on the east coast at the northern end of Anne Port Bay at La Crête Point (Nichols & Hill, 2004, p. 23) and north of Archirondel Tower to Le Malade at the southern end of St. Catherine's Bay. Rare examples have also been found by the author in the raised beach deposits in the second bay south of La Cotte de St. Brelade on the west side of Portelet Common headland

In terms of provenance, does this mean that these different pebbles from the east and south coasts have been transported and abraded by tidal currents and longshore drift driven by south - easterly winds during the sea - level rise over the last 10,000 years…leaving aside the possibilities of jettisoned ballast.

Yet another prospect is revealed among the granite bedrock which crops out through the sand further west near the end of the beach, and towards the back of this area there are two narrow, c. 1m wide dykes of grey dolerite rock, one of which bifurcates to form a third. They strike c. E - W and seem to have been unrecorded, possibly because they were formerly covered by sand (Fig. 15).

At the western end of the beach, past the houses, the yellow cliffs curve south

Fig. 15.

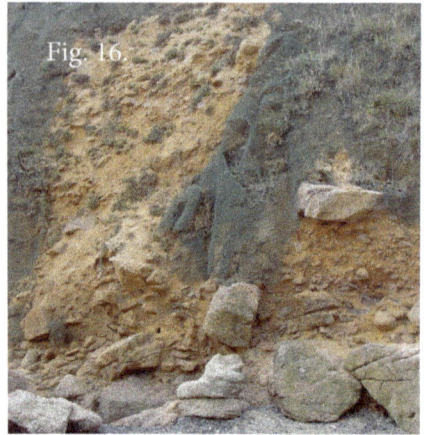

Fig. 16.

Fig. 17.

Fig. 18.

and are covered by green netting to stabilise the deposits and vegetation, but a partly broken section reveals thicknesses of angular blocks of granite forming head, and of yellow loess overlying a basal, thin strand of pebbles of granite and minor grey shale, and dolerite, from the dykes or sills which form the western part of the central raised beach (Fig. 16). Here, the raised beach section varies in thickness, related or controlled by the uneven nature of the granite bedrock. Some of the exposures are well cemented in parts (Figs. 17, 18).

Yet another prospect is gained by climbing carefully up onto the wave-cut platform which extends southwards below the former quarries in the western

Fig. 19.

Fig. 20.

cliffs. Here, there is a horizontal, grey - green dolerite sill, 30 - 60cm thick, partly forming the wave - cut groove (notch) where it has been eroded faster than the granite. About half way along, it has been cut and downthrown by an E - W dolerite dyke along a vertical fault (Figs. 19, 20). Other dykes also occur, eroded to form gullies and clefts, and looking back, eastwards across the bay, other narrow, vertical - sided, black gullies and clefts can be seen striking into the pink granite on the south side of l'Île au Guerdain and in the eastern cliffs in Fig. 2.

During an associated visit, our Chairman successfully located and photographed the dolerite sill at the base of the headland (Fig. 21), recorded on the IGS map but not previously published; she was unable to collect a sample due to the height of the tide.

Above the sill, the last prospect, there is one of the old quarries which reveal that the granite varies in texture and colour. In parts, there are porphyritic granites with large feldspar crystals in a finely crystalline groundmass (Fig. 22) and in others, more uniform textures or clusters of black biotite mica crystals (Fig. 23), a sample of which was collected by our Chairman for our rock database, and pegmatitic patches of milky quartz. Section members have also found similar varieties of granite and textures in the sea walls at La Haule and Millbrook which seem likely to have come from these quarries as they are nearer than those at La Rosière (La Moye).

Fig. 21

Fig. 22.

Fig. 23.

Brief Geological History.

During the Lower Palaeozoic, the Porphyritic granite (c. 550Ma) and the Aplite/microgranite (c. 527Ma) of the Southwest Granite Complex in the area described intruded the Precambrian Jersey Shale Formation after its deposition, uplift and folding, and during the eruption of the andesites forming the lower part of the Volcanic series (c. 522 – 477Ma). The adjacent Corbière granite was then intruded (c. 483Ma) and more andesites erupted, while during the same period (but later than 483Ma) there was intrusion of the dolerite sills then later, the dykes of the Jersey Main Dyke Swarm along E - W strikes.

Intrusion of the Northwest Granite Complex, occurred below, during the eruption of the rhyolites and the uplift and erosion and deposition of the Rozel Conglomerate (c. 477 – 426Ma).

The folding and intrusion occurred during a period from c. 700 – 425Ma, ie. c. 275Ma from the Precambrian to Silurian, a long period known as the Cadomian Orogeny (Brown et al, 1990, p. 181 et seq.).

This seems to have been followed by a long period of erosion, during the Upper Palaeozoic and Mesozoic, which removed the country rock and revealed the granites, until Tertiary limestones were deposited around the island.

During the Pleistocene there were several periods of higher sea level during interglacial times when raised beaches were formed and the raised beach and before deposits in Portelet Bay are representatives of the 8m one. These were interspersed with periods of loess and head deposition during intervening glacial times, the deposits in the present littoral zone were produced during the present sea - level rise. The present climatic regimes and weather, which control the present weathering, marine and fluvial erosion and deposition, have produced the present day features, for example, the tombolo spit linking the beach to l'Île au Guerdain.

References.

Bishop, A. C. & Bisson, G. 1989. Classical areas of British geology; Jersey: description of 1:250,000 Channel Islands Sheet 2 (1978). London HMSO for British Geological Survey.

Brown, M. Power, G. M. Topley, C. G. & R. S. D'Lemos, R. S. 1990. Cadomian magmatism in the North Armorican Massif. p. 181 - 213. In The Cadomian Orogeny. Eds. D'Lemos, R. S., Strachan, R. A. & Topley, C. G., 1990, Geological Society Special Publication No. 51. Geological Society, London.

Keen, D. H. 1978. The Pleistocene deposits of the Channel Islands. Rep. Inst.. Geol. Sci., No. 78/26.

Lees, G. J. 1990. The geochemical character of late Cadomian extensional magmatism in Jersey, Channel Islands, p. 273 - 291 in The Cadomian Orogeny, Geol. Soc. Spec. Publ. No. 51. (see D'Lemos et al. above).

Marett, R. R. 1911. Pleistocene Man in Jersey. Archaeologia, lxii, pp. 449-480.

Mourant, A. E. 1933. The Raised Beaches and Other Terraces of the Channel Islands. Geol. Mag. Vol. LXX, pp. 58-66.

Renouf, J. 1986. Geological setting and origin of La Cotte de St. Brelade, p. 35-52, in (Eds.) P. Callow & J. Cornford. La Cotte de St. Brelade: 1961-1978, Excavations by C.B.M. McBurney.

Renouf, J., James, L. High level shore features of Jersey (Channel Islands) and adjacent areas. Quaternary International (2010), doi: 10.1016/j.quaint.2010.07.005

Fig. 2.

Did you know...

That the overlying andesite volcanic rocks, seen at La Belle Hougue, are lava flows, ash and broken rock fragments (breccia) which form strato - volcanoes in the Andes today?

The Portelet Common coastal Trail.

Granites, rock shelter, raised beaches and dykes.

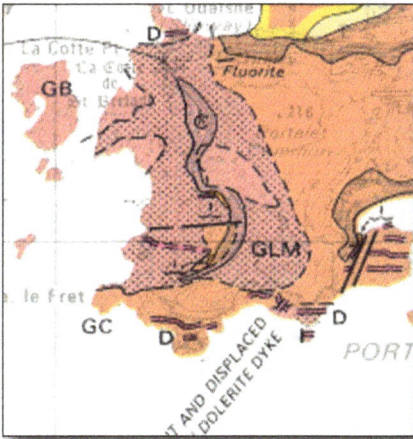

Fig. 1.

The Trail can start at L' Ouaisné beach and continue southwards via the quarries and the rock shelter, La Cotte de St. Brelade (Fig. 1), then along the beaches on the west side of Portelet Common (Fig. 2, facing page) and then back up onto Le Fret Point, but the main part of the Trail concerns the raised beaches, the trail for which starts from the west side of Portelet Common.

La Cotte is within the cliff outcrops (N) on the left hand side of the photo (Fig. 2), and Portelet Common forms the plateau surface along the top of the cliffs and slopes. The raised beach sites can be approached from the Portelet Common car park via the path southwestwards to a long granite wall, then west down the path across the heather - covered slopes to a stand of trees at the top of the low beach cliffs. From there, it is a short, steep descent via a path to the beach on the north side of the headland between the two bays. High tide times must be checked and care must be taken when walking over the boulders, rock ledges and low crags.

Many raised beaches at 8m, 18m and 30m around the island have been recorded but not always described, e.g. Marett (1911, pp. 449 - 480), Mourant (1933, p.60 - 61, Portelet to the east), Keen (1978), Renouf (1986, p. 39) and Bishop & Bisson (1989, p. 84, incl. Head, p. 87). Detailed work is at present in progress (Dr. J. Renouf, pers. comm.).

Fig. 3.

Fig. 4.

These raised beaches are in the 8m beach interval and are situated at the back of two bays, with present day pebble beaches, to the south of La Cotte de St. Brelade headland on the west side of Portelet Common facing St. Brelade's Bay.

These notes and illustrations help comparisons to be made between these and beaches in La Cotte de St. Brelade at the same elevation but they should be confirmed by detailed mapping and petrographical studies of the rock types.

In the first, smaller, northern bay to the left of the central crags (Fig. 2), accessible from a path down through several trees at its southern end, there are two possible raised beach levels. The first, starting from its first obvious appearance about half way northwards along the bay, consists of two possible pockets of pebbles which occur above Head; vegetation cover makes it difficult to trace any bedding but they seem to be part of a lower - level beach (Figs. 3, 4).

Above, is faintly stratified Head overlain by a second, seemingly higher pebble bed, 60 - 100cm thick, which is again overlain by more faintly stratified Head; the stratification being possibly due to solifluction or gelifluction (Fig. 5). Some metres further south, fine - grained, yellow sand and clay form a lens in the lower Head (Fig. 6).

Several metres further south (just north of the path descending onto the beach through the trees), it is thought that the upper pebble bed has dipped and thickened and lies in the grey soil profile within the first two metres above the present beach level (Figs. 7, 8).

Continuing southwards around the headland between the two bays, there is

Fig. 5.

Fig. 6.

Fig. 7.

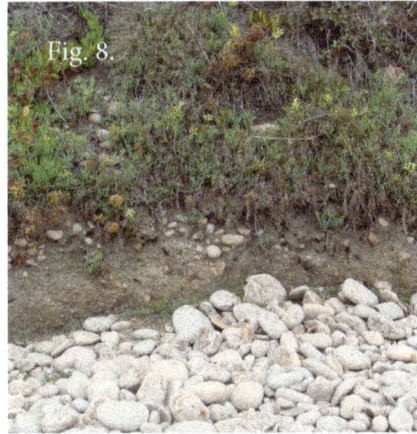
Fig. 8.

a small pocket of raised beach above the bedrock of porphyritic granite, the oldest of the South - west granite (Fig. 9). This raised beach deposit is immediately above a possible raised wave - cut platform at the foot of the headland and is noticeable for the smaller size of the pebbles (gravels) and their differing rock types, one of which is dark maroon and maybe rhyolite. Further on, round the south side of the headland, a narrow gully has been eroded along a discontinuous

Fig. 9.

dolerite dyke, c. 60cm wide, striking approximately E - W (Fig. 10).

In the second, southernmost bay, on the right of the central crags (Fig. 11), the raised beaches occur in the low cliffs at the back of the beach and extend southwards towards Le Fret Point.

They are separated from those of the northern bay by the small but high rocky headland of the SW granite complex, the central crags mentioned above (Fig. 11).

At about mid - tide level at the northern end, the beach is composed of beds of well cemented, light and dark coloured, generally small, rounded gravels, and scattered larger pebbles of many different rock types (Fig. 12).

Assuming this is a raised beach, an upper, second raised beach of larger, light coloured, rounded granite pebbles can be seen starting on the left (N) at c. 2 - 3m up the cliffs. This passes up southwards into an overlying thicker bed of brown, rounded pebbles which appears to descend southwards to beach level. The raised beach exposures are interbedded with intervals of yellow - light brown gritty sand, silty clay, and angular boulders in a yellow - brown sandy matrix which are thought to be loess and head.

The succession in detail is as follows;

Fig. 10.

Fig. 11.

Fig. 12.

Head; large angular boulders, 5 - 7cm to 15cm) towards top of cliff.

Sand; yellow – brown, with clay lenses, ?loess.

Pebbles of second raised beach; light grey & brown, granite, rounded, between angular boulders, and localised grey ?shale & dark grey red - spotted ?spherulitic rhyolite clasts.

Loess & Head; yellow to brown with some rounded grit clasts and angular boulders above.

Conglomerate of first possible raised beach. Small pebbles/gravel, 2 - 4cm, round, elongated, well - cemented; shale, granite, aplite, diorite, quartz and flint.

The lowest raised beach is at present day mid-beach level at the northern end of the second present day beach, and consists of c. 1m of small, 2 - 4cm rounded pebbles of a variety of rocks, eg. shale, granite, aplite, diorite, quartz and flint, generally in contact and well - cemented (Fig. 12). The matrix is brown - yellow sand and grit (Fig. 13).

Above is a layer of yellow - brown, medium - grained silty sand with some rounded grit, c. 60cm thick (Fig. 14), with scattered angular granite fragments of various sizes towards the top (Fig. 15). This is possibly loess overlain by head.

These deposits are overlain by the second raised beach, which can be seen starting higher and further left

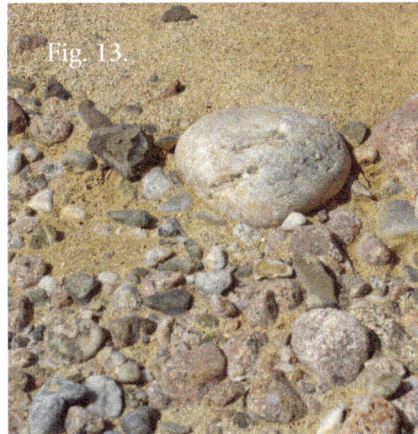

Fig. 13.

Fig. 14.

Fig. 15.

(N) of the lower one, above some possibly fallen angular blocks at the foot of the cliffs. It is a thin bed, c. 30cm thick, of clean, light grey, rounded granite pebbles, some of which occur in between large, angular, red granite boulders (Fig. 16). Further along it appears to thicken and pass upwards into a thick layer of larger, brown (possibly due to a sedimentary coating), well - rounded pebbles, seemingly off-set to the south (Fig. 17). This exposure and pebble colour may be the result of a cliff-fall and represent the internal unweathered part of the raised beach.

This in turn is overlain c. 1m of yellow sand with an upper layer of small angular rocks and possible loess.

Further south again, brown, silty clay lenses crop out in the beach below and

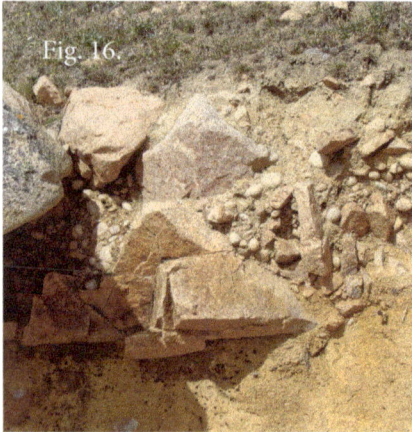
Fig. 16.

Fig. 17.

Fig. 18.

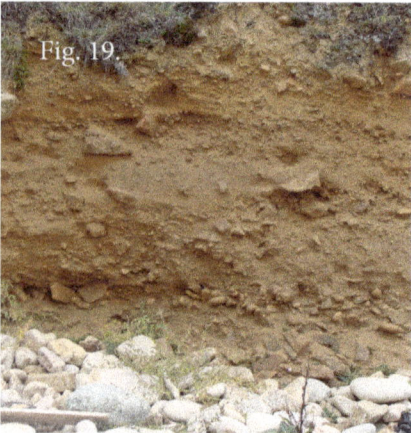
Fig. 19.

Fig. 20.

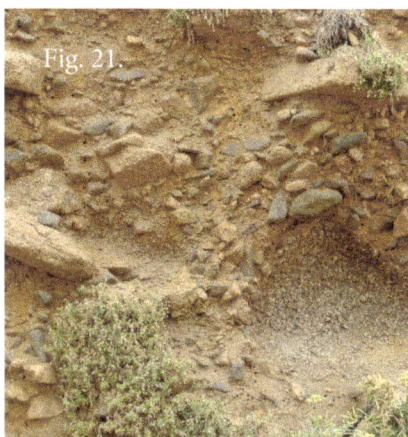
Fig. 21.

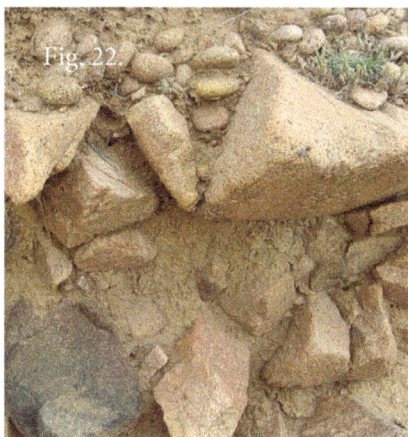
Fig. 22.

Fig. 23.

above the raised beach pebbles
(Figs. 18 & 19).

Within the raised beach here, there
are angular and rounded clasts that
superficially resemble dark grey
shale (Figs. 20, 21). These need to be
detached and examined closely as
they may possibly be dolerite from the
nearby dyke which could have been
eroded and provided beach pebbles.

Also, in this area of the cliff, just below
the bed of cleaner, smaller pebbles,
very unusual, larger cobbles of a red
'spotted' dark grey rock occur, not yet
seen in the bed rock of this area.

The red 'spots' are sub-spherical,
c. 1mm in diameter, and may be
composed of the red feldspar, common
in the underlying SW granite
(Figs. 22 & 23).

The shape and distribution of
the 'spots in this rock resemble
those in the variety of spherultic
rhyolite, composed of small, circular

spherulites, seen at La Crête Point on the east coast, north of Anne Port Bay. These need to be examined in more detail.

Finally, in the more inaccessible upper part of the cliff, angular boulders in the familiar yellow - brown ochre - coloured, fine sediment, possibly head (Fig. 24), overlie the pebble beds.

Ascending the raised beach cliffs back along the descent path, the trail continues to Le Fret Point and down the gently sloping cliffs, to examine the granite, its xenoliths of diorite and very rare laminated Jersey Shale, and the fine E - W striking dolerite dykes (Figs. 25, 26).

Brief Geological History.

The South-west Granite Complex consists of three granites of different ages, formed during the Lower Palaeozoic and which crop out in the SW corner of the island from La Corbière to Noirmont Point. Their outcrop patterns are oval and concentric about St. Brelade's Bay. The central area comprises a narrow band of Porphyritic Granite (La Moye granite) around a central Microgranite (Beauport or St. Brelade granite, see map and text, Bishop & Bisson, 1982 & 1989) and forms the oldest combination (c. 550 – 527 Ma); it forms the bedrock to the raised beaches described above. The outer Corbière granite is a coarsely crystalline granite

Fig. 24.

Fig. 25.

Fig. 26.

with hornblende and is the youngest, intruded c. 480 Ma.

The granites were intruded after the deposition, uplift and folding of the Jersey Shale Formation to the north and east and the intrusion of the diorites, xenoliths of which can be seen at Le Fret Point. This seems to have happened at about the same time (c. 522 – 477 Ma) as the andesites of the St. Saviour's Andesite Formation were forming to the north east which were then followed by the deposition of the Bouley Rhyolite Formation.

This was followed by the intrusion of the E - W dolerite dykes, part of the Jersey Main Dyke Swarm, seen along the Trail.

There seems to have been a very long period or periods of uplift and erosion while Tertiary limestones were being deposited on the sea bed around the island and before the formation of Jersey's raised beaches during the Pleistocene.

The loess deposits south of La Cotte near present beach level seem to represent deposition during a former glacial period, pre - Ipswichian/ Eemian, while the overlying raised beaches seem to be part of the 8m raised beaches which were deposited during the later Pleistocene sea level rise during the Ipswichian/Eemian c. 130 - 115,000 years ago. Discussion of the dating methods and results in the Jersey and the Channel Islands area is comprehensively presented by Renouf (2010, p. 1 - 4). The overlying head and loess represent a return to glacial conditions and gelifluction periods during the last glacial period, the Devensian, c. 115,000 - 10,000 yrs ago. The present sea - level rise and climatic regime during the last 10,000 years have produced the present landforms by marine erosion accompanied by physical and chemical weathering.

References.

Bishop, A. C. & Bisson, G. 1989. *Classical areas of British geology; Jersey: description of 1:250,000 Channel Islands Sheet 2. London HMSO for British Geological Survey, and Sheet 2, 1982.*

Brown, M. Power, G. M. Topley, C. G. & R. S. D'Lemos, R. S. 1990. *Cadomian magmatism in the North Armorican Massif. p. 181 - 213. In The Cadomian Orogeny. Eds. D'Lemos, R. S., Strachan, R. A. & Topley, C. G., 1990, Geological Society Special Publication No. 51. Geological Society, London.*

Keen, D. H. 1978. *The Pleistocene deposits of the Channel Islands. Rep. Inst.. Geol. Sci., No. 78/26.*

Lees, G. J. 1990. *The geochemical character of late Cadomian extensional magmatism in Jersey, Channel Islands, p. 273 - 291. In The Cadomian Orogeny, Geol. Soc. Spec. Publ. No. 51. (see D'Lemos et al. above).*

Marett, R. R. 1911. *Pleistocene Man in Jersey. Archaeologia, lxii, pp. 449-480.*

Mourant, A. E. 1933. *The Raised Beaches and Other Terraces of the Channel Islands. Geol. Mag. Vol. LXX, pp. 58-66.*

Mourant, A.E. 1935. *The Pleistocene deposits of Jersey. Bull.of the Soc.Jers.Vol.XII, pp.489-496*

Renouf, J. 1986. *Geological setting and origin of La Cotte de St. Brelade, p. 35-52, in (Eds.) P. Callow & J. Cornford. La Cotte de St. Brelade: 1961-1978, Excavations by C.B.M. McBurney.*

Renouf, J., James, L. (2010).*High level shore features of Jersey (Channel Islands) and adjacent areas. Quaternary International, doi: 10.1016/j.quaint.2010.07.005*

Did you know...

That the volcanic rocks above are rhyolites, seen at Bouley Bay and La Crête, a mixture of banded lava flows, flattened ash, pumice, breccia, striking spherulites and columnar - joints?

The Seymour Tower Trail.

Seymour Slipway - Tower; L' Angliaiche - Le Téton, and N to Le Hurel slipway; silts (?loess), sands, grits & littoral beach conglomerate.

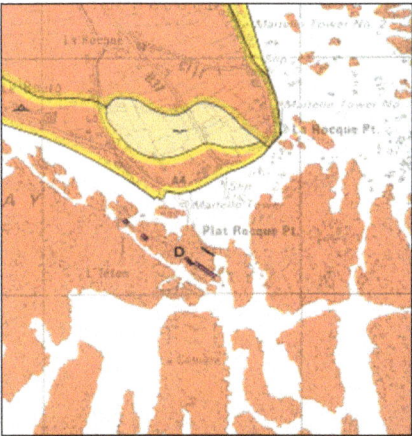

Fig. 1.

Silty clay forms horizontal to sub - horizontal deposits in hollows of various dimensions in the Southeast Granite Complex and is the most extensive deposit in the localities examined along and adjacent to the track (not marked) from the first slipway N of La Rocque (Fig. 1). It varies from grey to yellow - brown in colour and crops out as inliers

('windows') in the present day sands and gravels and in the pools of the granite wave - cut platform on either side of the white marker track from the slipway to the tower.

1. Seymour slipway to Seymour Tower.

Between Seymour slipway and Seymour Tower, via the Refuge (Fig. 2), the littoral zone or beach area is periodically covered and

Fig. 2.

Fig. 3.

Fig. 4.

Fig. 5.

uncovered by superficial, present day beach sand and grit deposits and so appears and disappears, but many have been GPS - plotted by Dr. A. Hill during the last few years (pers. comm.). Standing water pools in the south eastern area may indicate the presence of a more continuous layer underneath the beach sands, grits and pebbles, than at the other localities (Figs. 3, 4. 5).

2. L' Angliaiche to Le Téton.

The following deposits, found by La Société Jersiaise members Paul Chambers and Nick Jouault, are like the grey, but unlike the yellow and brown, clayey silts found by Arthur Hill and later by the author, near the Refuge further north between Seymour slipway and Seymour Tower.

Here, beds of light grey, brown - stained, consolidated silt and sand beds with scattered grit clasts and some quartz and rare flint crop out in the littoral zone SE of La Rocque harbour around the stacks named L' Angliaiche (Fig. 1, south centre) and Le Téton (Fig. 1, centre west). Above them, a thin outcrop of conglomerate of small to large, angular to rounded pebbles and cobbles in grey sand and silts form a separate outcrop and are thought to be part of a former beach conglomerate in the intertidal zone. In addition, a mica lamprophyre

Fig. 6.

Fig. 7.

Fig. 8.

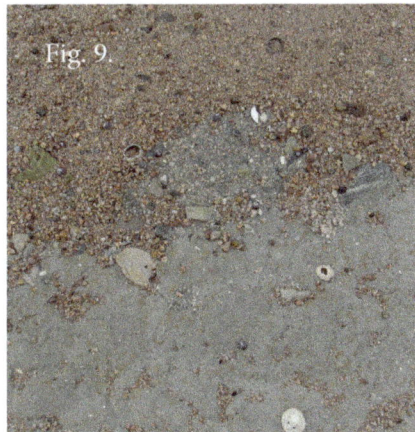

Fig. 9.

(minette) dyke and a wide quartz dyke, not recorded on the map, also crop out SE of the harbour.

The fine grained, grey - brown beds crop out around the base of L' Angliaiche (Fr. L' Anglaise), and lie on the granite bedrock, lapping around its uneven surface. The thickness is unknown and can't be determined as the base was not seen. In parts of the outcrop there are also brown patches or staining, indicating Fe precipitation between impressions of clasts removed by erosion (Fig. 6). The deposits are uniformly grey, fine - grained silt and fine sand, possibly with some clay. They show polygonal cracks, seemingly dessication cracks of different dimensions, and contain scattered gravel and rare cobbles of granite, quartz, rare flint between impressions of clasts presumably removed by erosion (Figs. 7, 8, 9).

The large clasts vary in colour, size and shape, from orange to grey, from

Fig. 10.

Fig. 11.

Fig. 12.

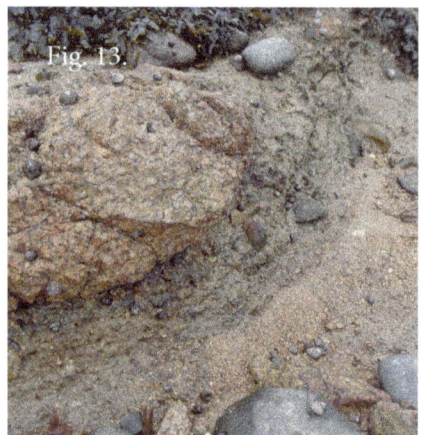
Fig. 13.

5 - 10cm, and from sub - angular to elongate. The rock types vary from granite to greywacke and banded diorite.

Littoral beach conglomerate.

Above this horizon, and further away, coarser, grey beds, 30 - 60cm thick, of poorly sorted, well - cemented pebbles and cobbles of varying size and shape, some angular, some long and thin with rounded edges (Fig. 10), and others, wedge - shaped with rounded edges in a finer, grey matrix (Fig. 11). One part of the outcrop consists of a thin wedge of finer grey sediments overlain by a wedge of a coarser, ferruginous, purple - coloured finer pebble bed under a granite boulder (Fig. 12), and another part consists of a large granite boulder lying in, and surrounded by the finer sediment which has been compressed around it (Fig. 13). The large pebbles and cobbles vary in colour from

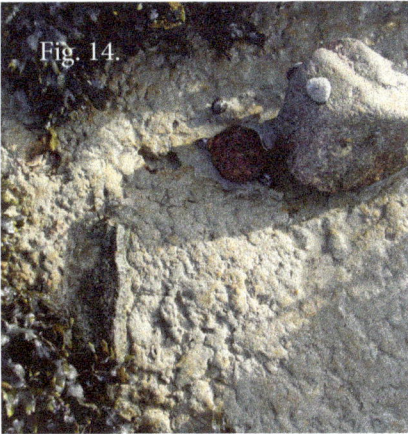
Fig. 14.

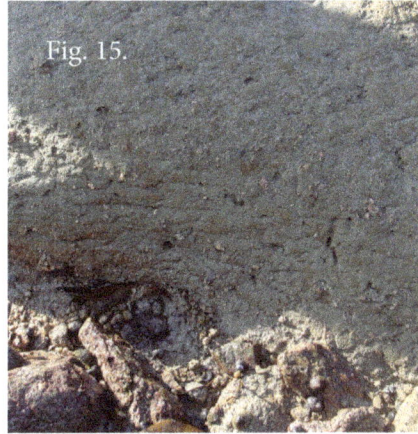
Fig. 15.

orange to grey and brown, in size from pebbles and cobbles,
10 - 15cm across, to small boulders, in shape from sub - circular pebbles to elongate and triangular large cobbles, and in rock type from orange, medium crystalline granite to greywacke and possibly maroon - brown rhyolite. Varying degrees of rounding and sphericity are also seen, from angular to sub - rounded and rounded, with low in the larger clasts to low sphericity in the pebbles. The smaller pebbles are often grain - on - grain supported in contrast to the larger ones which are often matrix - supported.

Further west across expanses of sand and small lagoons, is Le Téton, the highest part of a mass of low reefs and stacks SW of La Rocque harbour. A different deposit crops out between the reefs and low stacks. It is much thicker, with c. 1.5 - 2m showing above the general beach level and comprises a grey, compact, silty sand with grit - sized clasts in parts and larger pebbles in others (Fig. 14).

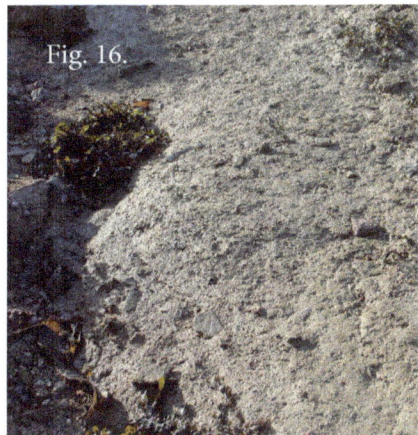
Fig. 16.

They are all matrix - supported and horizontal planes within the matrix make it appear thinly bedded in the lower part at the landward end of the outcrop (Fig. 15). The weathered surfaces are also uneven possibly due to weathering and/or shallow erosion along drying cracks. Their surfaces also exhibit a patchy brown iron staining. The outcrop becomes more uniform along the exposure, where the clasts become larger but still scattered within the enclosing sands and silts (Fig. 16). Continuing along the outcrop, it becomes thicker and is exposed below a covering of cobbles and boulders rising to a form the base of a boulder - strewn col between two stacks (Fig. 17).

To add to the variety of this area, there are two dykes, shown to the author by Paul Chambers, which crop out near

Fig. 17.

Fig. 18.

Fig. 19.

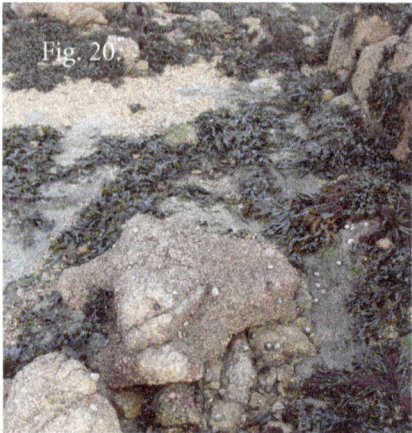

Fig. 20.

Fig. 21

L' Angliaiche, both unrecorded on the IGS map. One is made of white quartz c. 1m - 1.5m wide, and strikes c. WNW cropping out in the low cliff face of a small stack further west.

The second is a mica lamprophyre (minette) dyke, 1 - 1.5m wide, striking c. NNE. It crops out in a shallow gully eroded along it, and varies in colour from brown to grey and it also varies in shape and thickness; large brown to bronze biotite micas are easily visible and often give it a cleaved appearance (Figs. 18 - 21 photos by Paul Chambers).

3. North Le Hurel Slipway & recently exposed littoral clayey silts, sands & grits.

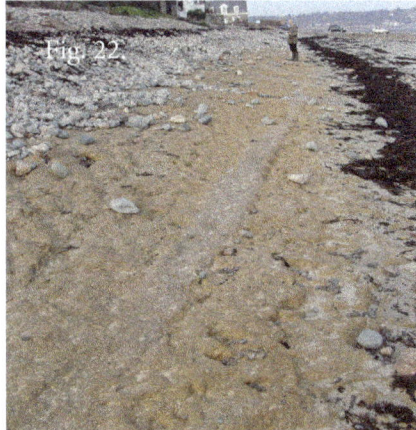

A further 500m north of the Seymour slipway, within 50m south of Le Hurel slipway, a longer outcrop of brown to yellow clayey silts was recently revealed at a higher elevation and occurs at the start of, and below, the high spring tide bank of pebbles against the sea wall (Fig. 22) reported by Dr. Paul Chambers (Dec. '09, pers. comm.).

They are light brown - yellow in colour, similar to some of the clayey silts further south near Seymour Tower, and seem to be the clay - silt with sand and grit variety, but in part with some black, possible carbon pieces (Fig. 23); more details will follow after examination by Dr. Chambers and Robert Waterhouse who also visited the exposure.

Pebbles resting on the surface seem to be sites of incipient pot - holing but in one case swirl marks occur in the sediments around a pebble clast (Fig. 24), while nearby, possible relict

Fig. 24.

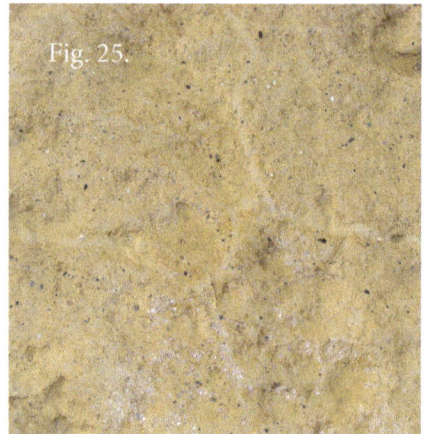
Fig. 25.

dessication cracks (first photographed by Dr. Chambers) stand out as light coloured lines within the yellow - brown background (Fig. 25).

Brief Geological History and possible correlations.

These above deposits, clay - silts and beach conglomerates in the littoral zone, lie in a similar position to those recorded from St. Brelade's Bay, La Cotte de St. Brelade – Le Fret Point, La Greve d'Azette, St. Catherine's Bay (south) and Bonne Nuit Bay.

They occur as three types of deposit, eg. semi-indurated, yellow to blue - grey and brown clayey silts (St. Brelade's Bay, Seymour Tower-Refuge); indurated, grey, yellow - brown, silts, sands, grits and pebbles (Le Hurel, St. Catherine's Bay south); and well - cemented, sometimes ferruginous conglomerates (La Cotte de St. Brelade – Le Fret Point, Bonne Nuit Bay).

The vari - coloured clayey silts are thought to be reworked loess, quite distinct from the loess above the 8m raised beaches in the cliff sections at the back of some of the bays.

The supposed reworked loess has not been analysed or examined for pollen so its redistribution mode, aeolian or fluvial/marine has not been determined.

The sandy grit and pebble beds may be a mixture of reworked loess and water - laid deposits of limited transport duration.

The well - cemented conglomerates (Mourant, 1933, p.59;1935, p.489; Keene,

1978b, p. 79; Bishop & Bisson, 1989, p. 85 - 86), quite distinct from the raised beaches in the cliffs, seem to represent former beach pebble - strand line deposits during the Pleistocene or Holocene as the sea level rose, with time for cementation which prevented their erosion during subsequent transgression to present levels.

Time intervals and their climatic variations have been established for the Holocene, above the Late Glacial of the Pleistocene, based on the Pollen Zones of Goodwin (1940), and the Pollen Assemblage Biozones & Radiocarbon start dates (bp) of Hibbert, Switsur & West (1971), in Jones et al (1990).

These are shown below and may be used to infer the conditions under which the sediments under discussion were deposited.

Pleistocene & Holocene subdivisions.

Interglacial		Holocene.	Flandrian (Chronozones, West, 1970).
Sub-Atlantic;	cold & wet.	VIII (Oak-Alder)	
Sub-Boreal;	warm & dry.	VIIb	5010 +/- 80 bp
Atlantic;	warm & wet.	VIIa (Oak, Elm, Alder)	7107+/- 120.
Boreal;	warm & dry.	VIc (Pine, Hazel, Elm)	8196+/- 150
		VIb (Hazel, Pine)	8880+/- 170
		VIa	
		V (Birch, Pine, Hazel)	9798+/- 200.
Pre –Boreal	sub-arctic	IV (Birch, Pine & Juniper)	c. 10,250.

Late Glacial		Pleistocene.	Devensian (Weichselian; Late Glacial).
Younger Dryas;	sub-arctic.		10,850.
Allerød oscillation;	warmer.		11,850.
Older Dryas;	sub-arctic.		12,050.
Bølling osc.	rapid warming;	sea level rise 100m	12,450. *Recent O isotope 14,600-14,100*
Oldest Dryas;	arctic.		15,000.
NB. Preceding Interglacial		130, 000 - 115, 000 Ipswichian (Eemian)	

Discussion

The various sediments occur midway between present day half - tide and MHWS levels lying unconformably on bedrock.

It is hoped that the period (s) of conglomerate and the possible redistributed loess deposition can be determined from the above climatic table, if not during one interval, then perhaps within a range of intervals.

Starting at the base, it is the thought the following is a possible sequence of conditions for our region.

From c. 130,000 – 115,000 years ago in the Pleistocene, during the last interglacial (Ipswichian (Eemian)), the sea bed around us, of Eocene limestone and Lower Palaeozoic rocks near Normandy and Les Minquiers, was covered by the sea which deposited the 8m raised beach.

Then, about 115,000 years ago, a late glacial period started (the Devensian (Weichselian)) and the sea regressed leaving us in a cold tundra type environment during the Oldest Dryas (a species of the Rosaecae arctic type of vegetation, with leaves similar to those of the oak).

During this time, the bedrock types would have been weathered physically by freeze - thaw and mass - wasting via solifluction and gelifluction, to produce layers of angular fragments comparable to head deposits and there would have been little or no deposition from rivers. Much airborne loess was probably deposited on and around the island during this time (Bishop & Bisson, 1989. p. 82).

This was followed between 14,000 - 12,000 years, by a rapid warming and transgression, the Bølling oscillation (named from Bølling Lake peat sequence in Jutland; pollen zone 1b), when the sea level rose 100m.

This level has been identified as still some distance offshore on maps by Lambeck (1995, p. 445).

Beach pebbles and gravels of the mixed rock types, (limestones, greywackes and quartz sandstones) would have formed at this level, while Temperate hardwood and softwood forests occurred on land between 29° - 41°N as the soil horizons developed, and streams and rivers would have increased flow and deposition onto the coastal areas and sea bed. The limestones would have been chemically weathered by solution.

From c. 12,000 – 10,000 there was a colder, sub-arctic period again, the Older Dryas, followed by a warmer period, the Allerød oscillation (named after the municipality of Allerød in Denmark; pollen zone II , mixed evergreen and deciduous), with present day - like temperatures, then followed by another colder, sub - arctic period, the Younger Dryas.

It is assumed that the sea - level fluctuated during these oscillations and that weathering and erosion of the bedrock occurred in similar relative ways, with more stream deposition during the warmer period, to give a relatively similar alternating sequence of sand, silt and clay sediments.

It is thought that the deposits examined could not have formed during this upper part of the Pleistocene, even though there would have been both terrestrial and marine redistribution, as subsequent sea - level rise during the Holocene (Recent) would have eroded them.

From 10,000 years ago to the present (Holocene (Recent)), starting with the Pre - Boreal, a sub - arctic period (? a continuation of the Younger Dryas), the climate became Boreal, gradually warmer and drier; then Atlantic, warmer and wetter; and lastly Sub - Boreal, warmer and drier again, until c. 5000 years ago. Sea temperatures were also rising with different flora and fauna appearing, while on land, different vegetation types were colonising the area.

NB. It was during the Atlantic period that the peat beds were deposited.

This is also the date (c. 7,000 bp / 5,000BC) generally given for our low - lying plateau areas and crags to have become islands and reefs.

Consequently, it is in this interval, possibly at the end of the warm and dry Sub - Boreal period, that the climatic conditions existed for the formation of the cemented beach conglomerates and their part ferruginisation with the Fe being provided because of the temperature and humidity conditions, at a relevant sea - level height in the present day intertidal zone, comparable elsewhere (see below).

It is also possible that the warmth and dryness created an arid enough environment for aeolian action to redeposit the loess. One assumption is that, as the land surface was soil - covered, the cliff deposits and stream bank deposits were the source of the loess and that it was re - deposited by wind action. An alternative assumption is that streams transported it to the sea where longshore drift redistributed it in hollows and protected areas of the bedrock.

Subsequently, until today, the climate became colder and wetter and is now called the Sub - Atlantic!

The identification of the floral changes through pollen analysis has been done, but a palynologist is still needed to examine and date the shoreline sediments described above, and more importantly, to examine the core sediments across the adjacent channel, La Déroute.

References.

Allaby, A. and Allaby, M. 1990. A Dictionary of Earth Sciences. 2nd. Ed. Oxford University Press.

Bishop, A. C. & Bisson, G. 1989. Classical areas of British Geology: Jersey: description of 1:25,000 Channel Islands Sheet 2. (London HMSO for British Geological Survey).

Jones, R. L., Keen, D. H., Birnie, J. F. & Waton, P. V. 1990. Past Landscapes of Jersey. Société Jersiaise, p. 3.

Keen, D. H. 1975. Two aspects of the last interglacial in Jersey. Ann. Bull. Soc. Jersiaise. Vol. 21. pp. 392 - 396.

Keen, D. H. 1978b. The Pleistocene deposits of the Channel Islands. Rep. Inst. Geol. Sci. No. 78/26.

Lambeck, K. 1995. Late Devensian and Holocene shorelines from the British Isles and North Sea from models of glacio - hydro - isostatic rebound. Quart. Jour. Geol. Soc. v.152, p. 437 - 448.

Mourant, A. E. 1933. The raised beaches and other terraces of the Channel Islands. Geol. Mag., Vol. 70, pp. 58 - 66.

Mourant, A. E. 1935. The Pleistocene deposits of Jersey. Ann. Bull. Soc. Jersiaise. Vol. 12, pp. 489 - 496.

Roberts, H. M. 2008. The development and application of luminescence dating to loess deposits: a perspective on the past, present and future. Boreas, Vol. 37, No. 4. pp. 483 - 507. Wiley - Blackwell.

Wintle, A. G. 2008. Luminescence dating: where it has been and where it is going. Boreas, Vol. 37, No. 4. pp. 471 - 482. Wiley-Blackwell.

GLOSSARY

Agglomerate. Volcanic rock consisting of angular volcanic fragments and sometimes of other rock types, and different sizes, and shapes set in fine-grained solidified volcanic ash.

Andesite. A finely crystalline, igneous volcanic rock of intermediate composition, a mixture of iron and magnesium minerals and feldspars.

Ankerite. A calcium, magnesium, manganese carbonate mineral. It is closely related to dolomite in its composition.

Biotite. A black to brown flaky mineral, in a subgroup of the Mica group.

Boudinage. Sausage, lens or rectangular - shaped structures due to attenuation from differential load pressure in semi - consolidated beds.

Breccia. Rock made of coarse - grained, angular fragments of rock in a matrix of finer material.

Clast. A rock fragment or grain resulting from the breakdown of any rocks.

Col. The lowest point or pass between two hills or peaks.

Columnar jointing. A series of generally hexagonal columns formed by vertical joints as a result of contraction during the cooling of a lava flow.

Concretion. A small, hard, compact mass of sedimentary rock formed by the concentration of minerals within the spaces between the sediment grains; usually spherical.

Conglomerate. A rock made of rounded pebbles or cobbles in a matrix of smaller grains.

Diorite. A coarse - crystalline, intrusive (plutonic) igneous rock of intermediate composition.

Dolerite. A dark - coloured igneous rock with medium sized crystals, usually forming dykes and sills.

Dolomite. A carbonate mineral and rock composed of calcium magnesium carbonate.

Dyke. A narrow wall - like structure of igneous rock that formed at mid-crustal levels in joints or faults in a pre-existing rock body.

Eutaxitic. A banded appearance in hardened volcanic ash, resulting from the layering of glass shards and pumice (silica).

Fault. A fracture in the rock where adjacent sides have been displaced.

Feldspar. An alkali silicate mineral, a most common igneous rock-forming mineral.

Felsite. A crystalline igneous rock made of quartz and feldspar; sometimes containing larger crystals and forming dykes and veins.

Fiamme. Pumice clasts inside an ignimbrite that have become stretched sideways while still hot as the deposit is compressed by the weight of overlying material.

Flow-banded rhyolite. A rhyolite flow that shows flow-banded textures formed as the lava flowed down the volcano.

Flow-banding. A structure that shows banded textures, formed as the lava flowed down the flanks of a volcano.

Galena. The natural mineral form of lead sulfide. It is an abundant and widely distributed sulfide mineral and is the most important lead ore mineral.

Gabbro. A dark-coloured plutonic intrusive igneous rock made of large crystals of iron and magnesium minerals.

Granite. A light-coloured plutonic intrusive igneous rock made of large crystals of feldspar and quartz minerals.

Granophyre. A variety of granite that contains intergrowths of quartz and feldspar.

Greywacke. A sandstone of quartz, feldspar and rock fragments (clasts) in a matrix of clay and silt.

Haematite. The most important ore mineral of iron. It is an oxide mineral that is found in a variety of colours and forms.

Head. Describes overlying glacial deposits (of sand, silt and angular rock fragments) at the very top of the cliffs, produced by soil flow after frozen ground has thawed.

Hornfels. A finely crystalline metamorphic rock with the hardness and texture of an animal horn, produced by thermal metamorphism.

Hornblende. A complex series of minerals. Not a mineral in its own right but is a general term to refer to dark amphibole.

Ignimbrite. A volcanic ash deposit consisting of fragments lava, and pumice which turns to glass when welded together by high temperatures to become banded in the lower part.

Intrusion. An igneous rock that has been formed under the Earth's surface from magma which has slowly pushed up into any cracks or spaces it can find.

Ipswichian Interglacial. The second to latest interglacial period of the last ice age between 130,000 to 110,000 years ago.

Iron pyrite. Also known as 'fool's gold', this is an iron sulfide made of iron and sulphur.

Joint. A fracture in the rock without any displacement of the adjacent sides, in contrast to a fault.

Lamprophyre. A dark - coloured dyke rock with large crystals surrounded by small crystals. Mica lamprophyre contains minerals from the Mica group.

Loess. A fine, silt - sized sediment which is formed by the deposition of windblown silt and clay.

Micas. A group of minerals with a sheet silicate structure, biotite and muscovite being the most common.

Microgranite. A medium - crystalline intrusive igneous rock. It contains crystals which are smaller and more uniform than those of granite, indicating that the magma has cooled more rapidly.

Molybdenite. A mineral made of molybdenum and sulphur, a molybdenum sulphide.

Mudstone. A fine - grained sedimentary rock made of clay or mud.

Parataxitic. A exceptionally well - layered and drawn - out eutaxitic texture.

Periglacial. Refers to places around the edges of glaciated areas.

Playa. A clay pan, mud flat, or dried up lake.

Pleistocene. The geological Epoch which lasted from 1.6 million to 11,700 years ago, spanning the world's recent period of repeated glaciations and interglacials.

Pluton. A body of intrusive igneous rock that has crystallized from magma cooling deep below the surface of the Earth (a plutonic rock).

Porphyry. A variety of igneous rock consisting of large crystals dispersed in a finely crystalline matrix.

Quartz. An abundant mineral in many igneous, sedimentary and metamorphic rocks, and made of silicon and oxygen

Quaternary Period. A geological time Period between 1.6 million and the present day, consisting of the Pleistocene and Holocene Epochs,and is characterized by a series of glaciations and interglacials and the appearance of modern humans.

Rheomorphic. Used to describe flow structures that are only observed in high grade ignimbrites.

Rhyolite. A light coloured volcanic rock, rich in silica, and made up of small crystals. The extrusive equivalent of granite.

Shale. A fine - grained, flakey (fissile) sedimentary rock made of mud that is a mix of flakes of clay minerals and tiny fragments of other minerals.

Siltstone. A fine - grained sedimentary rock made of silt-sized particles; coarser than shale and finer than sandstone.

Slump folds. These occur when layers of sediment move down a slope and fold or crumple.

Soil creep. The movement of soil down a slope under gravity.

Sphalerite. A mineral that is the main ore of zinc. It is usually found in association with galena, pyrite and other sulfides; sometimes referred to as zinc blende.

Spherulites. Small, usually spherical bodies consisting of radiating crystals and curved (concentric) bands, found in silica - rich volcanic rocks.

Stalagmite. A conical pillar in a limestone cave that is gradually built upward from the floor as a deposit from ground water seeping through and dripping from the cave's roof.

Strike - slip fault. A fault with a near - vertical plane where the adjacent walls move to the left or right past each other horizontally, with very little vertical motion.

Tombolo. A depositional landform in which an island is attached to the mainland by a narrow piece of land.

Turbidity current. A rapidly flowing current, laden with sediment, flowing down a deltaic or continental slope.

Wolstonian. A middle Pleistocene glacial stage between 352,000 and 130,000 years ago that preceded the Ipswichian interglacial.

Zinc blende. See Sphalerite.

Acknowledgements

A sincere thank you is given to all those who have helped to interpret the geology that is seen along the trails. Over the years, Dr. Arthur Mourant, Dr. John Renouf and Dr. Michael Andrews have all encouraged the studies and freely discussed their findings which they have published in many papers. The authors of the texts consulted, such as Clive Bishop, Gary Bisson and Stan Salmon, in addition to Arthur Mourant and John Renouf and Mike Andrews have also given their valuable time to discuss elements of their research, which gave further valuable insight into the interpretation of Jersey's fascinating and unique geology. In addition, members of the Geology Section of La Société Jersiaise under the chairmanships of Dr. Arthur Hill, Warren Hobbs, Sandra Mahé and Colin Cheney have contributed to the following descriptions through their many questions and discussions, and the present author then revisited all the sites to re-examine them and provide more detailed descriptions and the photographs.

A particular thank you goes to Mark Jackson for editing and designing this book and for his excellent illustrations. I am similarly grateful to Paul Chambers for his advice and assistance with getting this and other books into print.

All the photographs reproduced in this publication were taken by the author. The map extracts were created by the British Geological Survey and are reproduced under the terms of Crown Copyright. The illustration of Ralph Nichols was created by Mark Jackson.

Information about the geology along the Trails is also available from the excellent BGS Channel Islands Sheet 2, Jersey (1978) 1:25,000 geology map, and the BGS Bishop, A.C. & Bisson, G. 1989, Description of 1:25,000 Channel Islands Sheet 2.

About the author

Ralph A. H. Nichols, Ph.D. (Wales), formerly worked as a research geologist doing field mapping, well site geology and laboratory work in Australia and Canada. Subsequently he became a Lecturer and Teacher of Geology and Geography in Canada and Jersey and is currently a member of La Société Jersiaise working in the Geology, Archaeology and Jèrriais Sections.

www.ingramcontent.com/pod-product-compliance
Lightning Source LLC
Chambersburg PA
CBHW042117190326
41519CB00030B/7529